Nuclear Engineering Division 2016

AA008960

Core Programming Area at the 2016 AIChE Annual Meeting

San Francisco, California, USA
13 - 18 November 2016

ISBN: 978-1-5108-3443-9

Printed from e-media with permission by:

Curran Associates, Inc.
57 Morehouse Lane
Red Hook, NY 12571

Printed by Curran Associates, Inc. (2017)

For permission requests, please contact AIChE
at the address below.

AIChE
120 Wall Street, FL 23
New York, NY 10005-4020

Phone: (800) 242-4363
Fax: (203) 775-5177

www.aiche.org

Additional copies of this publication are available from:

Curran Associates, Inc.
57 Morehouse Lane
Red Hook, NY 12571 USA
Phone: 845-758-0400
Fax: 845-758-2633
Email: curran@proceedings.com
Web: www.proceedings.com

TABLE OF CONTENTS

476892 Overview of U.S. DOE Environmental and Waste Cleanup Programs

Abstract:

The Department of Energy has a liability for the safe cleanup of the environmental legacy brought about from decades of nuclear weapons development and government-sponsored nuclear energy research. The Departmentâ€™s Office of Environmental Management (EM) is responsible for this cleanup function and over the 25-plus years of its existence have safely cleaned up 91 sites, leaving 16 sites, with these remaining 16 sites posing the greatest technical and programmatic challenges. Four of the remaining 16 sites are tank waste sites where wastes resulting from reprocessing of spent fuel are stored in over 200 large underground tanks, containing approximately 100 gallons of highly radioactive and chemically toxic wastes. EM has a nominal $6 Billion annual budget with approximately 35% of the budget targeted to safe management, treatment and disposition of tank waste as well as closure of the tanks

The four sites include the Savannah River Site, Hanford, the Idaho National Laboratory (INL) and the West Valley Demonstration Project (WVDP). Progress in waste retrieval, treatment and disposition vary among the sites, with the treatment mission essentially complete at WVDP and not yet substantially started at INL and Hanford. To accomplish this tank waste cleanup mission, EM and its field sites utilize approximately 5,000 contractors employed by various corporations, usually in the form of limited liability partnerships. In addition, EM and its contractors rely on specific technical expertise from national laboratories, outside technical experts, and universities to provide assistance. This is particularly true in the technology development area, in which partnerships with other agencies, e.g., NASA for robotics, and university partnerships are being utilized. This talk will address these aspects that are enablers to achieving the EM tank waste mission.

475885 Perspectives on Chemical Engineering Research in the Nuclear Fuel Cycle

Abstract:

This talk will focus on the history and role of Chemical Engineers at INL, with emphasis on the areas where I have worked, including areas that involve collaboration with other labs. I will also discuss the opportunities, challenges and outlook in the nuclear industry for current and future chemical engineers.

456816 Predicting Chromatographic Separations of Lanthanide Chelates for Nuclear Forensics

Abstract:

In an effort to meet the growing demand for more rapid separation methods, this work reports the development and thermodynamic characterizations of organometallic species as a vehicle for the rapid separation of volatile nuclear fission products. Volatile organometallic complexes can be rapidly separated from each other via gas phase chromatographic separations due to differences in adsorption enthalpy. Because adsorption and sublimation thermodynamics are linearly correlated, there is considerable motivation to determine sublimation enthalpies to improve the ability to predict separations effectiveness. A method of using thermogravimetric analysis (TGA) is employed in this study on thirteen lanthanide 1,1,1,5,5,5-hexafluoroacetylacetone (hfac) complexes ($Ln[hfac]_4$). Sublimation enthalpies are determined using the Langmuir method involving isothermal jump heating. An empirical correlation is used to estimate adsorption enthalpies of lanthanide complexes on a quartz column from the sublimation data. TGA-MS and melting point analysis are used to confirm sublimation as well as confirm the stability of the chelate through the sublimation region. Finally, a Monte Carlo model based on single-atom chemistry used for simulating separations on a thermochromatographic column is briefly described.

453056 Preparation of Surrogate Post-Detonation Debris Using a Plasma Torch

Abstract:

Preparation of Surrogate Post-detonation Debris Using a Plasma Torch

P. A. Taylor, A. D. Braatz and A. V. Giminaro

A commercial plasma spray system is being used to prepare powders that have been exposed to the extremely high temperatures present in a plasma torch (10,000–15,000°C). The feed powders represent materials present at the Trinity site in New Mexico and at selective urban sites[1]. The goal of the project is to determine the impact of the high temperatures on particle dispersion and leaching rates. The primary constituents of the feed are silica (~60 wt%) and alumina (~14 wt%), with smaller amounts of calcium oxide (~9 wt%), potassium oxide (~6 wt%), sodium oxide (~2 wt%) and magnesium oxide (~2 wt%). The urban feed has an iron concentration similar to the calcium, plus trace amounts of many other metals. The powder feeder for the plasma gun can handle particle sizes ranging from 20–200 µm, but a narrow particle size distribution is needed to ensure that the particles are uniformly exposed to the plasma arc. A laser-based particle visualization system will be used to show the path of the particles into the plasma arc. The flow rate of the carrier gas can be adjusted to optimize the exposure of the particles to the arc. The average surface temperature of the particles inside the plasma arc will also be measured. A spray dryer will be used to produce composite particles of uniform size, using small amounts of an organic binder. Early tests of the system used a Clinoptilolite zeolite powder that had been sieved to a 106–150 µm particle size. The zeolite has a similar composition to the urban sites. Figure 1 shows the plasma torch in operation, and Fig. 2 shows representative zeolite particles before and after exposure to the torch.

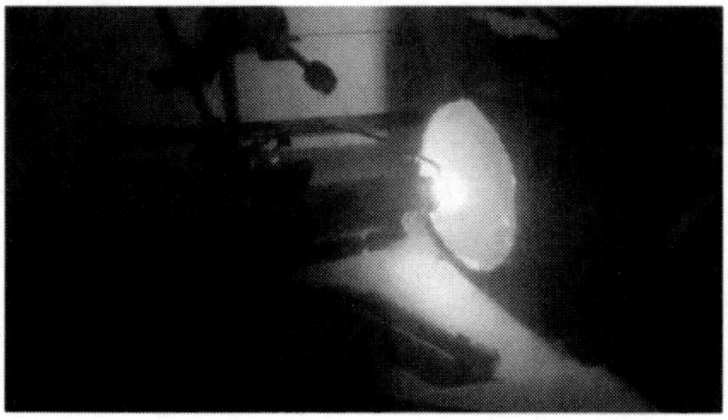

Fig. 1. Plasma torch viewed through a welding screen; plasma gas of 9% hydrogen in argon

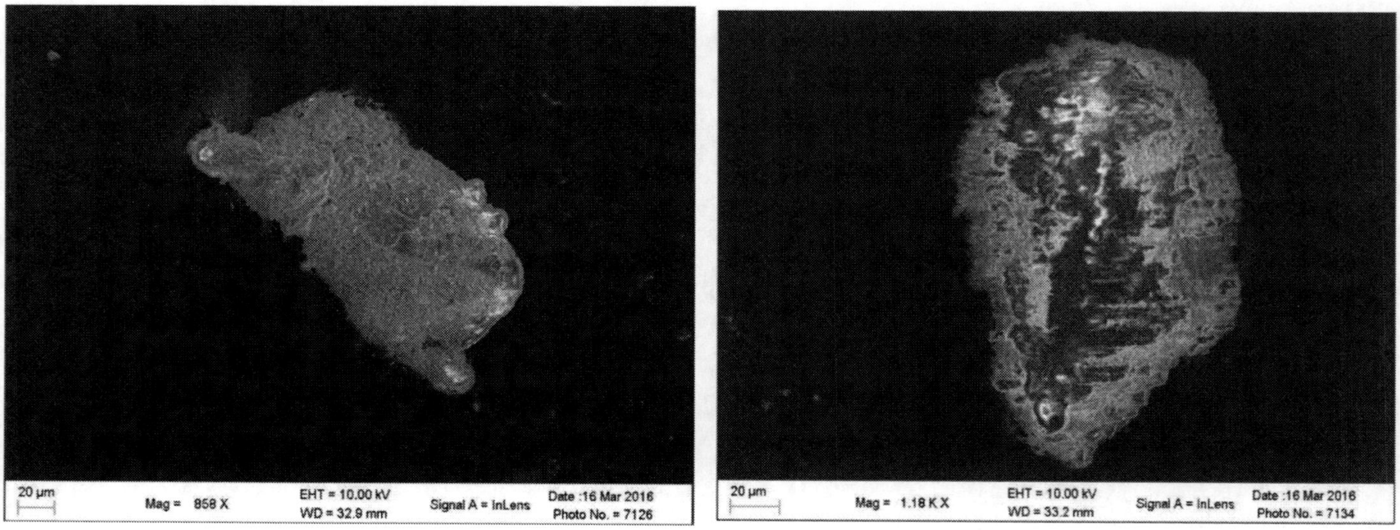

Fig. 2. Scanning electron microscope photo of as-received zeolite particle (left) and a particle after exposure to the plasma torch

1. A. V. Giminaro et al., "Compositional planning for development of synthetic urban nuclear melt glass," J Radioanal Nucl Chem, 306:175–181, 2015.

474398 Experimental Analysis of Plutonium Product and Raffinate Streams from Multiple Purex Process on a Low Burn-up, Fast Neutron Irradiated DUO_2 Pellet

Abstract:

Experimental investigations of separating actinides (uranium and plutonium) from fission products (FPs) were conducted using modified Plutonium Uranium Recovery by Extraction (PUREX) process. The samples under investigation were from a low burn-up (< 5 GWd/tU), depleted uranium dioxide (DUO_2) pellet (~13 milligrams) irradiated in a fast neutron spectrum in the High Flux Isotope Reactor (HFIR) at Oak Ridge National Laboratory (ORNL). This sample simulated the DUO_2 blanket region of a Fast Breeder Reactor (FBR) which was removed at low burn-up for weapons-grade plutonium (WGPu) recovery. The potential to misuse the material generated in this type of reactor prompts interest in a detailed analysis of the FPs present in the separated plutonium product and raffinate streams. A detailed understanding of these unique compositions would aid nuclear forensics activities for source attribution in the case of smuggled plutonium interdiction or at an inspection of a reprocessing plant to catch the misuse of such a plant. Aliquots of the dissolved neutron irradiated DUO_2 sample containing FPs and WGPu, underwent three modified bench-top scale PUREX processes. Each process had its own specific goal; maximize Pu recovery; minimize FP contamination in the product Pu; and characterize Pu decontamination factor (DF) values for a single stage extraction and back-extraction. Through these processes each step was analyzed via high-resolution gamma spectroscopy or ICP-MS. Distribution ratios and DF values were calculated for nuclides of interest to be incorporated into a larger nuclear forensics database. The reported experiments were performed in an argon atmosphere glove box, with low oxygen, moisture concentrations, and at room temperature.

A MECHANISTIC INTERPRETATION OF ACTIVITY PICK-UP DATA

James T. Orr
Naval Nuclear Laboratory
Schenectady, NY

Biography

The author is a senior advisory scientist at the Naval Nuclear Laboratory in Schenectady, NY. He received a PhD in Chemistry from the University of California at Berkeley in 1993 and an SB in Chemistry from the Massachusetts Institute of Technology in 1986.

Abstract

A series of long-term activity pick-up tests was conducted to assess the tendency of various materials to pick up radioactivity upon exposure to coolant of a pressurized water reactor (PWR). Testing was conducted at the Advanced Test Reactor (ATR). Several materials with a variety of surface treatments were included in the tests. A review of the gamma-ray spectroscopy data has identified three different mechanisms by which radioactivity is picked up: incorporation of radionuclides into a growing corrosion film, direct activation of elements within an alloy and non-specific adsorption of radioactive particulates. This paper provides the data analysis which identifies these distinct mechanisms by which radioactivity is picked up. This methodology may prove useful in assessing or predicting activity pick-up in other materials or in other systems.

Introduction

Radioactivity levels and the distribution of radioactivity are an important consideration in the commercial nuclear industry [1]. Management of personnel dose rates affect routine maintenance plans and can be quite costly. In general, one would like activated radionuclides to accumulate in purification systems or in well-shielded regions of the plant, and not accumulate in areas that require more frequent maintenance or are in close proximity to such areas. As a result, minimizing deposition of radionuclides is one of the design considerations in nuclear plants.

The Naval Nuclear Laboratories has performed a series of long-term activity pick-up tests at the Advanced Technology Reactor (ATR), a DOE facility in Idaho. Test specimens (flat coupons) of various materials were inserted into holders which went into simulated reactor coolant in various loops of the ATR. Exposures lasted up to 5.5 years, with intermittent examinations. Gamma-ray spectroscopy was performed on the test specimens, and the activities of 19 radionuclides were recorded. In all cases to be discussed, four replicate specimens of each material were included. The materials tested here were 304 stainless steel (UNS S30403), Ni-Cr-Fe Alloy 600 (UNS N06600), Ni-Cr-Fe Alloy 718 (UNS 07718) and Ni-Cr-Mo Alloy 625 (UNS N06625). Some specimens were also electropolished to identify the effect of electropolishing on activity pick-up.

The test program was concluded in 2005. The focus of the testing was to compare the tendency for these materials, whether in the as-machined or electropolished condition, to pick up radioactivity. The results presented here are the result of a subsequent review of the data. As such, there was no opportunity to direct the experiments to test the hypotheses presented herein. This report provides data analysis directed toward identifying the mechanisms of activity pick-up in the test specimens. Recognition of the relevant mechanisms of activity pick-up may improve the ability to predict activity distribution in a plant of interest.

Methods

The observed activities of 19 radionuclides, in μCi/specimen, were recorded for each test specimen. The activity of a radionuclide in a specimen was compared to the corresponding activity of the same radionuclide in a different specimen with the same exposure in the ATR to determine what effects were significant in the resulting activity. Three mechanisms of activity pick-up were considered: non-specific adsorption, direct activation and incorporation into a growing corrosion film.

Non-specific adsorption

In the case of non-specific adsorption, a radioactive particulate drops out of the coolant and lands on a surface. This is governed by the local flow characteristics of the circuit, and therefore the nature of the surface, or test specimen, is irrelevant. If one were to plot the activity of each nuclide on a specimen of a given material, A, versus the corresponding activities on a specimen of material B in the same test, one would expect the data to result in a straight line of unit slope. The activity of a certain radionuclide on each specimen is determined by its abundance and its specific activity in the coolant. There would naturally be some variability in the observations, but there would be no effect of the nature of the specimen material. This situation is depicted in Figure 1 below.

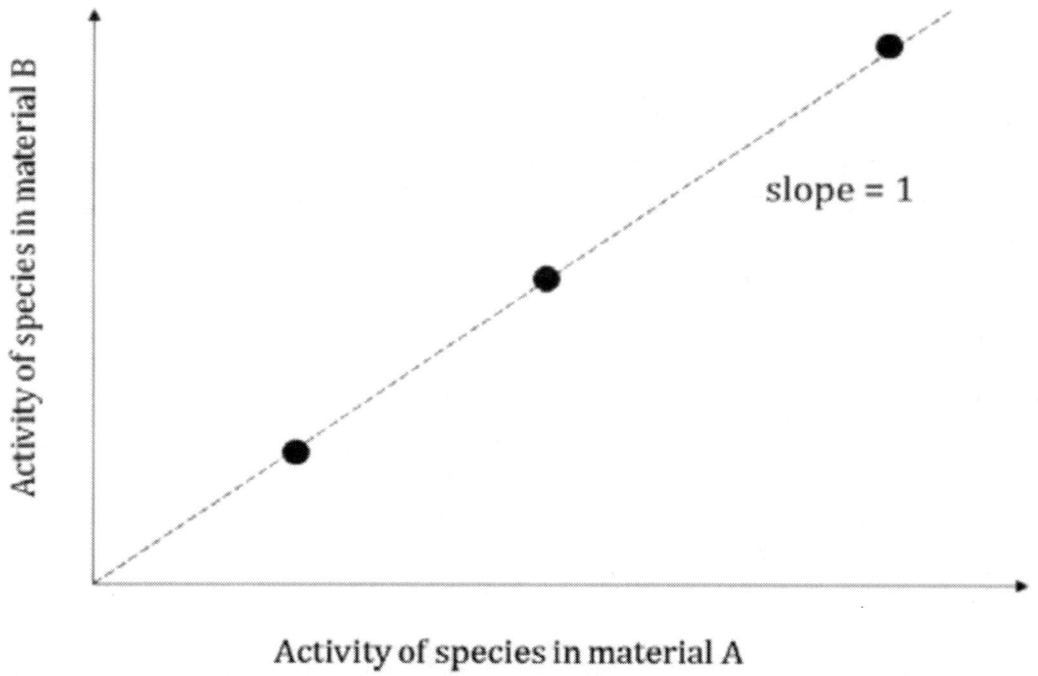

Figure 1. Depiction of non-specific adsorption mechanism, in which activated particles come to rest on specimens regardless of their composition.

Direct activation

In the case of direct activation, an element in the substrate material is activated directly by the neutron flux. This is highly dependent on the nature of the element, as each element has its own neutron absorption (or "capture") cross section, and the proximity of the specimen to the flux field. When comparing the activity of a specific radionuclide ("X") activated directly in two specimens of different materials, the material with more X in it will have proportionately greater activity of X. The intensity of the flux field is also a factor. This situation is depicted in Figure 2 below, assuming two specimens are at approximately the same flux intensity.

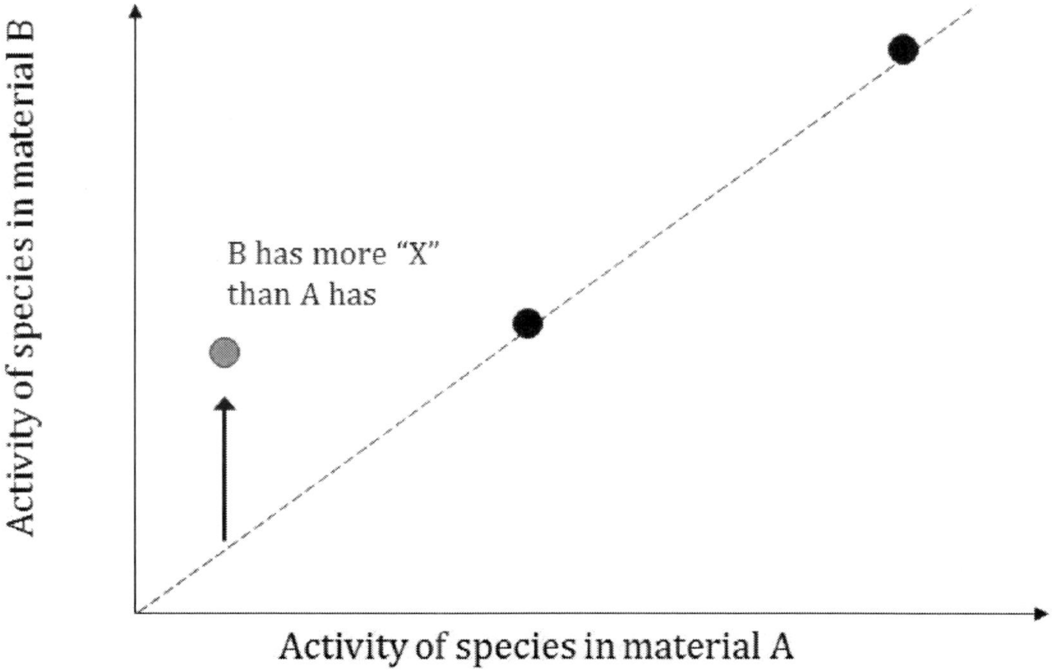

Figure 2. Depiction of direct activation mechanism, in which elements in the specimen material can be activated based on their cross sections and proximity to the neutron flux.

Incorporation into a growing corrosion film

It is well known that metal cations dissolved in the aqueous phase can become incorporated into corrosion films [2]. If some of these dissolved metal cations are activated, such as ^{60}Co, the substrate will become radioactive as the Co(II) is incorporated into the corrosion film. This is a mechanism by which activity can be deposited on surfaces far from the nuclear core. Ions in solution pass through the core region and become activated, and travel with the coolant through the circuit, eventually depositing elsewhere. Of particular interest are ^{60}Co and ^{58}Co because they have relatively long half-lives.

Corrosion films on stainless steels and many Ni-based alloys tend to be comprised of metal oxides of the spinel type, particularly ferrites (MFe_2O_4) and chromites (MCr_2O_4) [3-5]. In these oxides, there is a single divalent cation (M = Fe^{+2}, Co^{+2}, Ni^{+2}, etc.) and two trivalent cations (Fe^{+3} or Cr^{+3}). There have been many studies on the thermodynamics of these compounds [6-8].

When incorporation into a corrosion film is the dominant mechanism of activity pick-up, one expects to see activities of each radionuclide higher in the faster corroding alloy than in the slower corroding alloy, in proportion to the corrosion rates. A given radionuclide may have high or low activity, based on its concentration in the coolant and its nuclear cross section, but it will be higher in the faster corroding alloy since there are more sites to occupy in the thicker corrosion film. This situation is depicted in Figure 3 below.

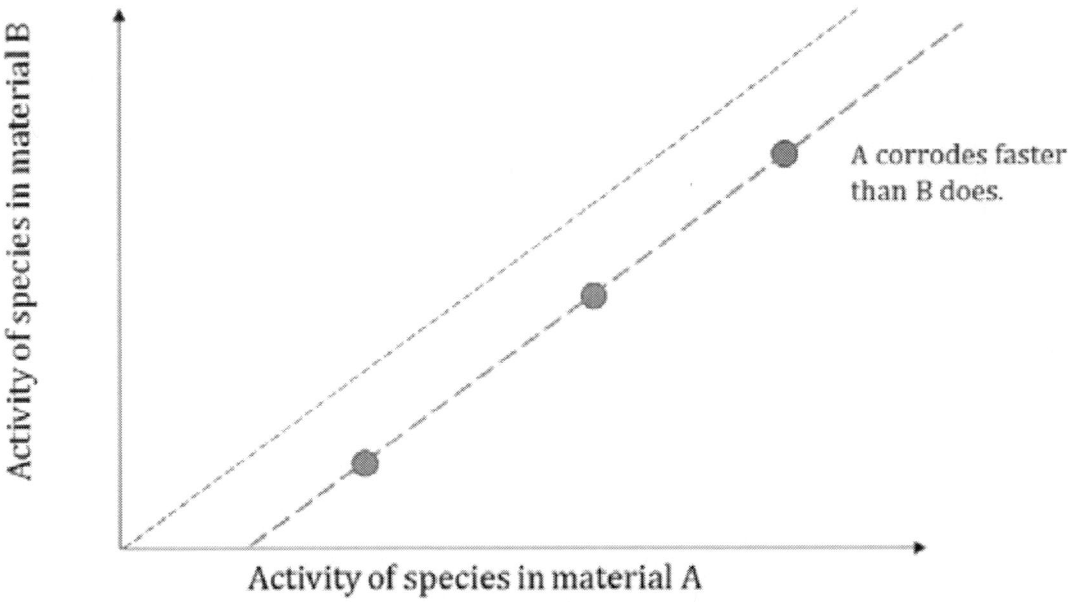

Figure 3. Depiction of incorporation mechanism, in which radionuclides dissolved in the coolant are incorporated into corrosion films of the specimen material.

These mechanisms of activity pick-up are specific to the various radionuclides. Within each specimen, all three mechanisms may be evident for different radionuclides. Comparison of activities of each radionuclide on a specimen to the corresponding activities on another specimen of a different material or condition is helpful to identify which mechanisms are consistent for the various radionuclides.

Results

The gamma-ray spectroscopy data from test specimens in two long-term exposure tests were analyzed according to the method outlined above. One test, run in the northwest (NW) loop of ATR, had a maximum exposure time of 49,770 hours above 400°F. The other test, run in the southwest (SW) loop, had a maximum exposure times of 28,793 hours above 400°F, although the data analyzed were taken from the 25,188-hour interim examination. (The final data were analyzed long enough after removal that radioactive decay made much of the data useless.) The SW loop of ATR had Zn added to the simulated coolant. Many commercial nuclear plants use Zn addition to the reactor coolant to decrease radiation levels [9].

In Figure 4 below, the activities of 19 radionuclides on Alloy 600 (A600) specimens are compared to the corresponding activities on 304 stainless steel (304SS). The radionuclides with the greatest activities (^{60}Co, ^{58}Co, ^{51}Cr and ^{95}Nb) are highlighted. Both the Alloy 600 and the 304SS specimens were in the as-machined condition, and received a 16-day pre-treatment in high-temperature water to begin establishing a corrosion film prior to exposure in the ATR. In all such figures in this report, the line of unit slope is included for reference.

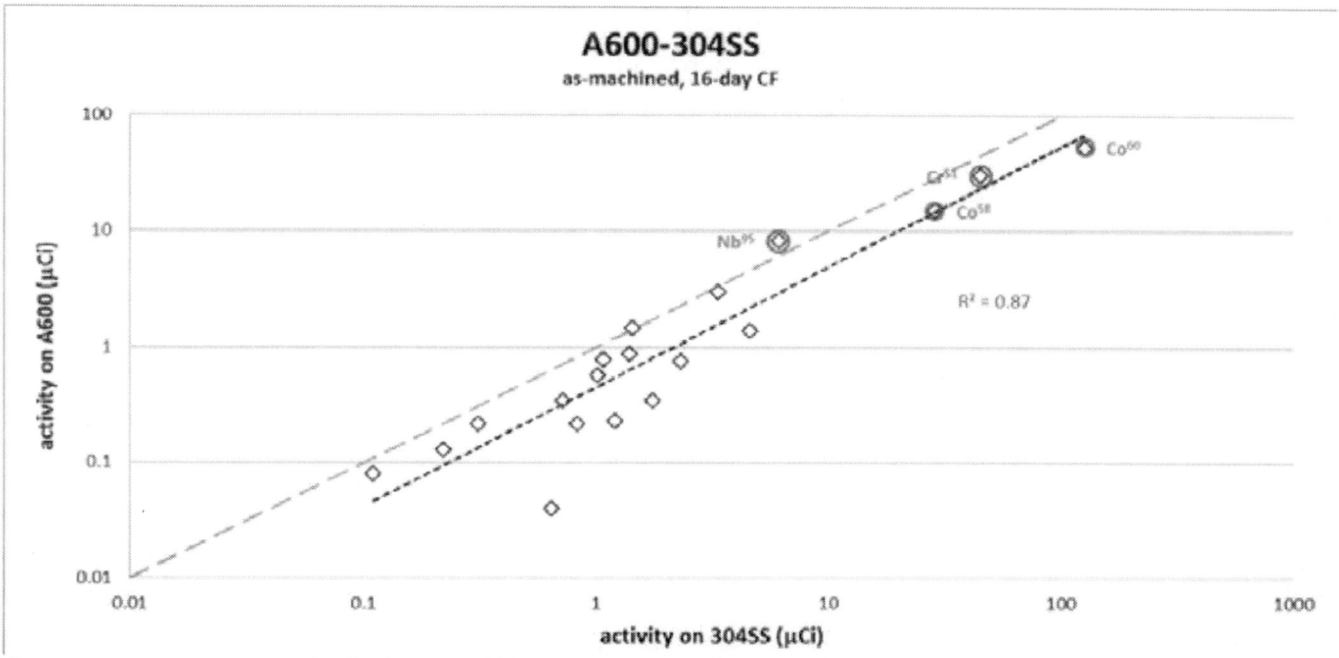

Figure 4. Comparison of radionuclide activities on Alloy 600 specimens to the corresponding activities in 304SS specimens exposed in the same test. The line of unit slope (dashed) is shown, along with a best-fit line through the data ($R^2 = 0.87$).

In Figure 4 there is clearly a bias in the direction of higher activities on 304SS, which corrodes faster in high-temperature water than does Alloy 600 [5, 10]. This suggests that the majority of the activity pick-up on these specimens was due to incorporation of radionuclides into the corrosion films. The best-fit line drawn through the data in Figure 4 shows a slope very nearly equal to unity (1.039), as would be expected if the same mechanism were responsible for the activity on both materials. This does not imply that all of the nuclides shown were affected by the incorporation mechanism, but that incorporation appears to be the dominant mechanism.

Of the radionuclides of highest activity, a notable outlier is ^{95}Nb, which appears to have roughly equal activity on Alloy 600 and 304SS specimens. This species is unlikely to be incorporated into corrosion films, as Nb is a +5 cation, and therefore unlikely to be a good fit in the spinel structure.

The activity of ^{51}Cr is somewhat higher on 304SS, but does not follow the trend line of all radionuclides. It is not clear from this figure what the behavior of ^{51}Cr is, but it is noted that 304SS has about 20% more Cr than does Alloy 600. Also, ^{51}Cr had unusually high specimen-to-specimen variability in all of the materials, as will be discussed later. It is further noted that Cr has very low solubility in high-temperature water in the absence of oxygen.

Figure 5 below presents a similar comparison between Ni-Cr-Mo Alloy 625 and 304SS. All specimens were as-machined and had a 16-day pre-exposure corrosion film.

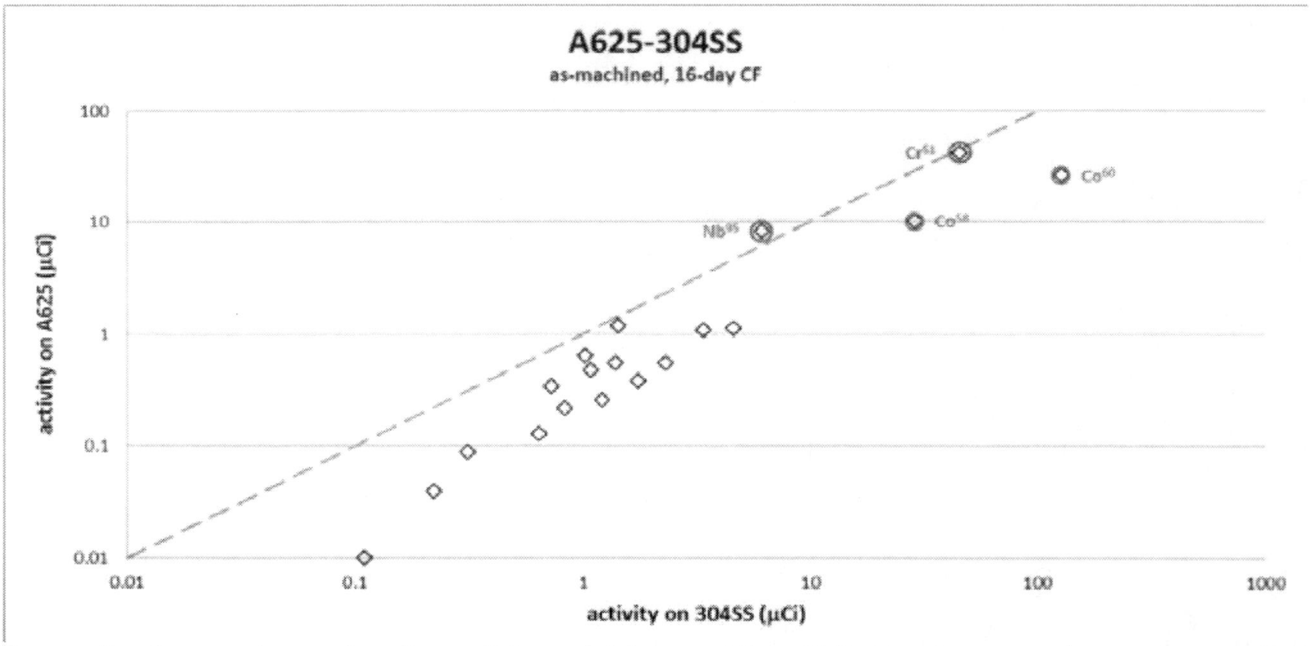

Figure 5. Comparison of radionuclide activities on Alloy 625 specimens to the corresponding activities in 304SS specimens exposed in the same test. The line of unit slope is included.

The corrosion rate of Alloy 625 in high-temperature water is lower than that of 304SS, so the same trend observed in the Alloy 600 example (Figure 4) is evident in Figure 5: the majority of radionuclides have higher activity on the 304SS specimens than on the Alloy 625 specimens. This is clearly the case with ^{60}Co and ^{58}Co, which are known to become incorporated into corrosion films.

The activities of ^{95}Nb and ^{51}Cr are both very close to the line of unit slope. In the case of Nb, this is an interesting observation since Alloy 625 contains approximately 3 wt% Nb, whereas the composition of 304SS specifies <0.1 wt% Nb. Although the abundance of Nb in Alloy 625 is more than an order of magnitude greater, its activity is virtually the same as that in 304SS. This strongly suggests that direct activation is not a significant factor in the observed activity of ^{95}Nb in the test specimens. There is only about 25% more Cr in Alloy 625 than there is in 304SS on a relative basis, so it is not possible to distinguish between the effects of direct activation and non-specific adsorption from this example.

The activities on electropolished and as-machined Alloy 718 are compared in Figure 6 below. Electropolishing is an effective means of reducing the corrosion rate of many stainless steels and Ni-based alloys, but it has a negligible effect on the bulk composition of the material. There are some who report Cr-enrichment as a result of electropolishing, but this appears to be a near-surface effect, and not one that would alter the composition of a test specimen to any noticeable degree. Therefore, comparisons like that shown in the figure below emphasize the impact of the incorporation mechanism.

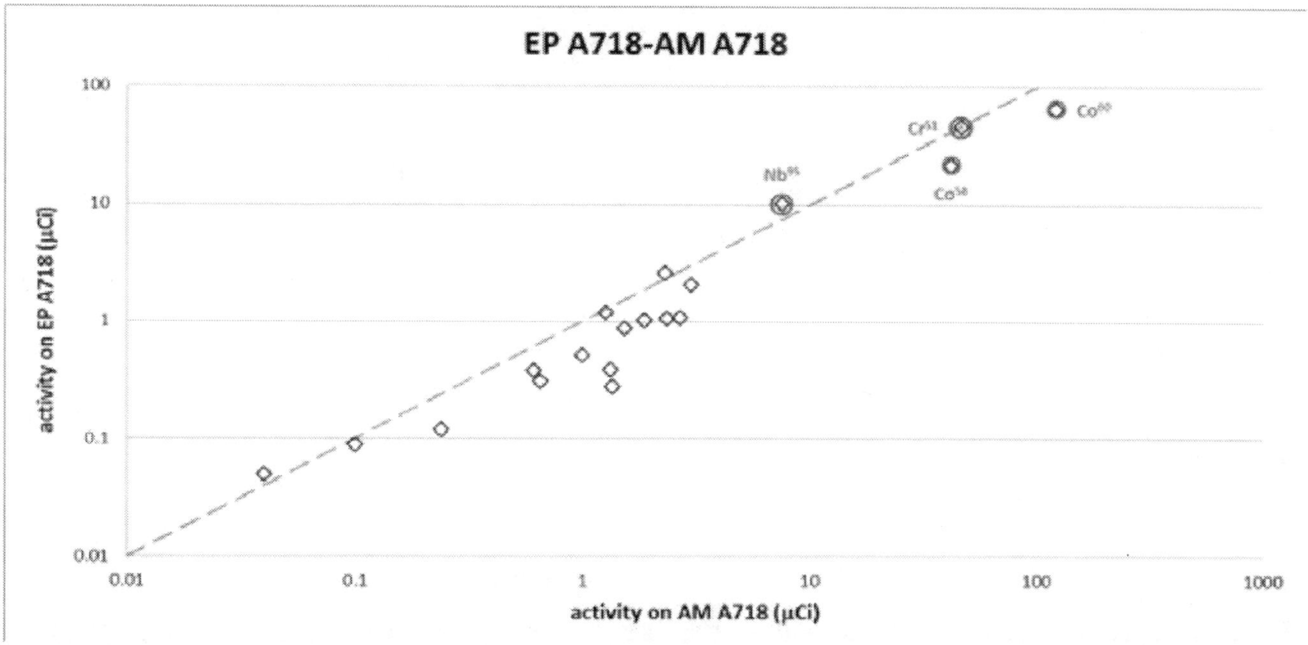

Figure 6. Comparison of radionuclide activities on electropolished Alloy 718 specimens to the corresponding activities on as-machined Alloy 718 specimens exposed in the same test. The line of unit slope is included.

The results shown in Figure 6 above demonstrate that most of the radionuclides investigated show greater activity on the faster-corroding material, which in this case is as-machined Alloy 718. This is consistent with the conclusions drawn from the previous two figures. The two notable exceptions, ^{95}Nb and ^{51}Cr, are the same two radionuclides identified in the previous figures. Both of these are very close to the line of unit slope in Figure 6. This could signify that their activities are proportional to their abundance in the material or that they are the result of non-specific adsorption, in which the base material plays no role.

All of the results presented so far have been from the 49,000-hour exposure in the NW loop. Figure 7 below shows results from the 25,000-hour exposure in the SW loop, in which the simulated coolant contained deliberately-added Zn. It is a comparison of activities on Alloy 625 and 304SS. The test specimens were in the as-machined condition with no pre-exposure corrosion film.

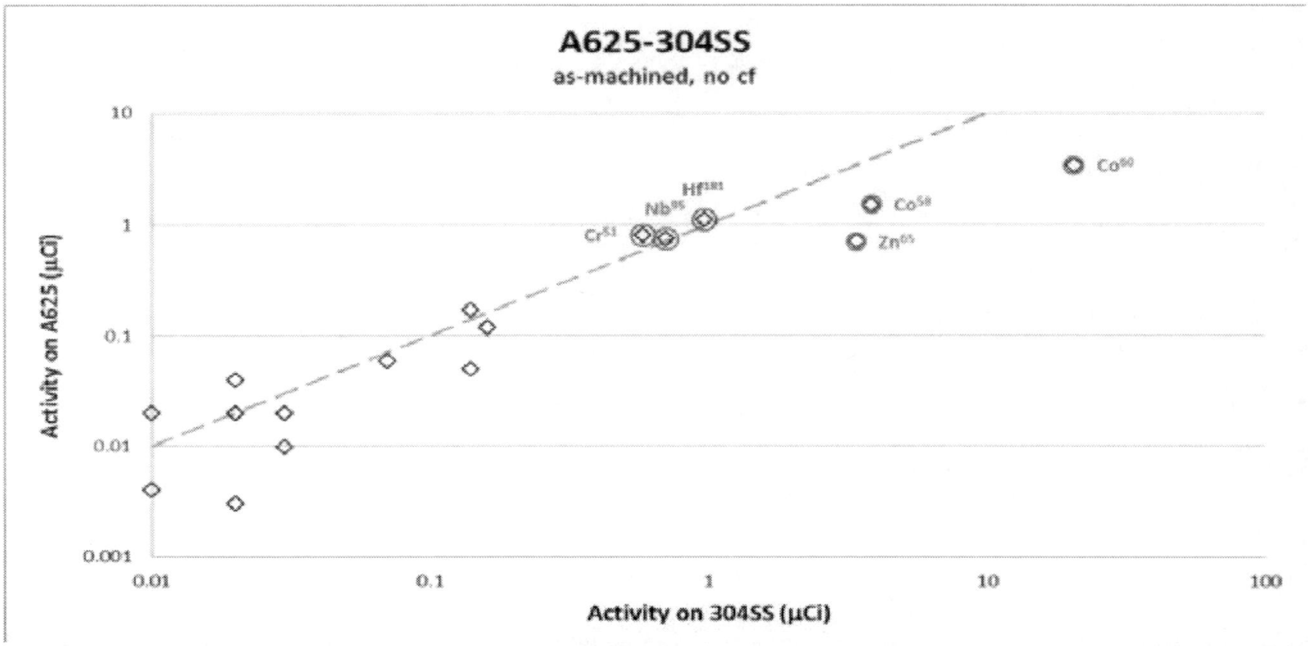

Figure 7. Comparison of radionuclide activities on Alloy 625 specimens to the corresponding activities on 304SS from the test in the SW loop, which contained added Zn. The line of unit slope is included.

Figure 7 shows some significant differences from the comparison of the same materials in Figure 5 above. Although the activities in general are much lower than they were in Figure 5, from the loop without Zn addition, one must bear in mind that the specimens in the SW loop had a much shorter exposure time, and the proximity of the samples to the flux field may have been different.

More significantly, one can see that the activities of ^{65}Zn and ^{181}Hf are much more prominent than they were in the non-Zn exposures. The increase in Zn activity is due to the presence of Zn in the simulated coolant. As expected, it appears to demonstrate behavior consistent with incorporation into a corrosion film, as it has much higher activity on the faster-corroding 304SS. Its behavior is similar to that of ^{60}Co and ^{58}Co.

The activity of ^{181}Hf falls very close to the line of unit slope, similar to the behavior of ^{95}Nb and ^{51}Cr. The source of Hf is believed to be test specimens of other tests being run at the same time in the same loop by other experimenters. Somehow the presence of Zn in the aqueous phase caused Hf to become more mobile, or perhaps to become more likely to be activated.

Thus far a clear distinction has been shown between the behavior of radionuclides whose activities are affected primarily by the corrosion rate of the test specimen and those that are largely unaffected by the corrosion rate of the test specimen. In the former category are ^{60}Co, ^{58}Co, ^{59}Fe, ^{54}Mn and ^{65}Zn. Several of these transition metals are known from the literature to become incorporated from the aqueous phase into corrosion films. In the latter category are ^{51}Cr, ^{95}Nb, ^{95}Zr and ^{181}Hf. This categorization of the radionuclides is consistent for all materials tested, whether as-machined or electropolished.

The data presented thus far have not allowed any distinction to be made between the other two postulated mechanisms, non-specific adsorption and direct activation. This is the subject of the remainder of the data analyses.

As stated earlier, if direct activation is the applicable mechanism of activity pick-up, one would expect to see an effect of the amount of the activatable species in the material. The average activity of ^{51}Cr on as-machined test specimens of the four materials is plotted below in Figure 8 as a function of the amount of Cr in the materials (in wt%). The amount of Cr varied over a small range for these materials.

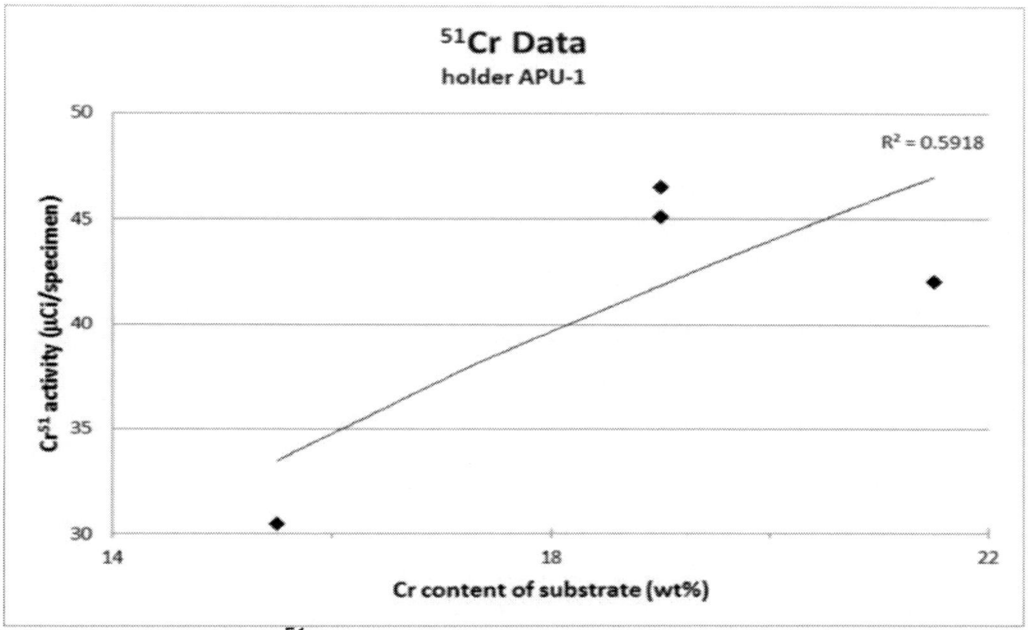

Figure 8. Average activity of ^{51}Cr on as-machined test specimens as a function of the weight percent of Cr in the material, with the best-fit line (R^2 = 0.59).

Although this plot appears to show a trend with the amount of Cr in the material, which would be consistent with the direct activation mechanism, it does not show the standard deviation of the plotted points. Each point is the average of measurements on four replicate specimens, and the standard deviation in all cases is greater than 50%. This is one of the respects in which the behavior of ^{51}Cr is unlike that of the other radionuclides that are unaffected by the material corrosion rate: the others have lower sample-to-sample variability. Figure 9 below shows the individual ^{51}Cr data points of all four materials.

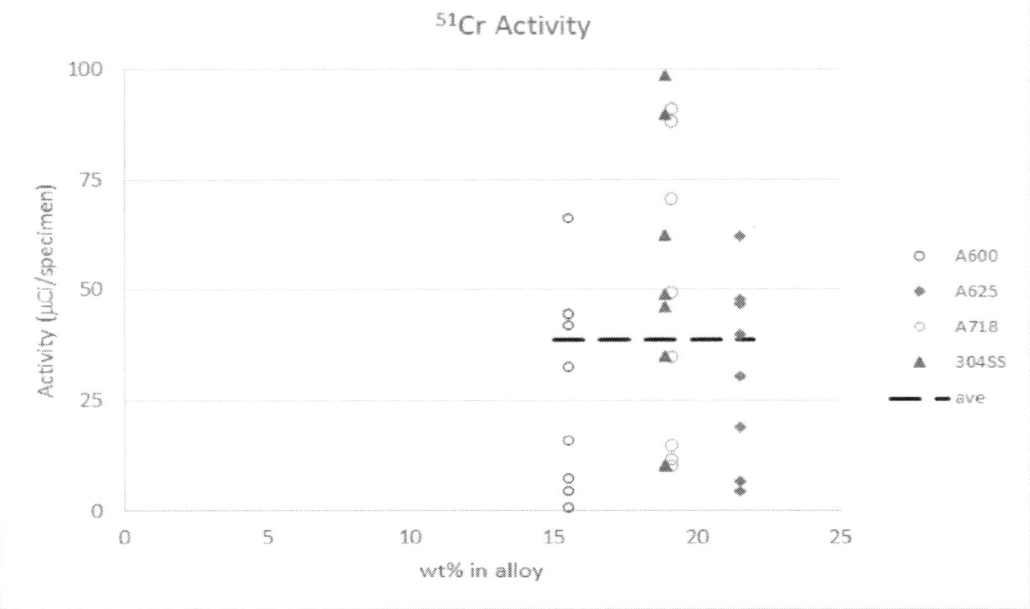

Figure 9. Activity of ^{51}Cr on all specimens as a function of the weight percent of Cr in the material. The average activity is shown, as the slope was not significantly different from 0.

In Figure 9, there is no trend of increasing ^{51}Cr activity with increasing amount of Cr in the material. The unusually high variability in the ^{51}Cr measurements makes rigorous comparisons more difficult. A similar plot for ^{95}Nb activity showed no dependence on the amount of Nb in the materials, which varied by more than an order of magnitude (<0.1 to 3 wt%). The specimen-to-specimen variability for ^{95}Nb was lower than that for ^{51}Cr, both in absolute magnitude and as a percentage of the average value.

Part of the variability in the ^{51}Cr observations was found to be due to the location of the specimens within the sample holders. The sample holders contained four rows in which the specimens were placed, and records of the specific location of each sample were retrieved. The observed activities of ^{51}Cr in a given material for samples in row A, which was farthest from the flux, were smaller than the corresponding activities of ^{51}Cr in the same material for samples in the middle of the holder (row B). This comparison blocks for the amount of Cr in the material, the ease with which Cr is activated and the surface condition of the sample. The same comparison was made for the other radionuclides which were unaffected by the material corrosion rates (^{95}Nb, ^{95}Zr and ^{181}Hf). Figure 10 below shows the results in terms of the difference in activity between a certain radionuclide on a certain material for a specimen in row A and the same radionuclide on the same material for a specimen in row B. To normalize the results, the differences were divided by the average activity for the radionuclide on all specimens of that material. This provided the differences on a consistent scale to facilitate comparisons.

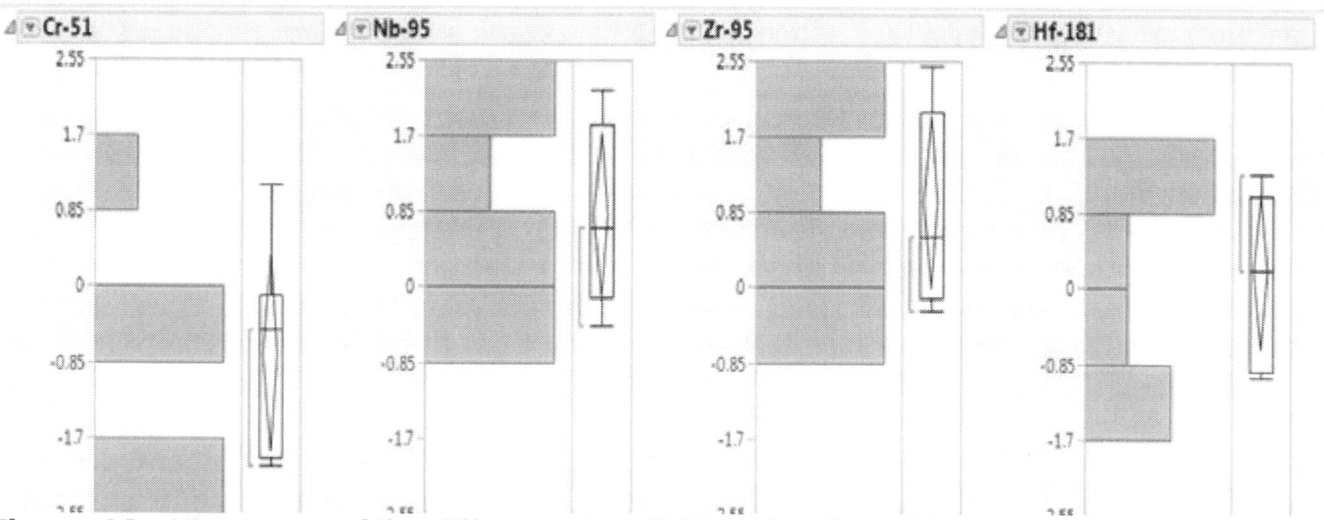

Figure 10. Histograms of the difference in activities of radionuclides on specimens in row A of the holder (farthest from the flux) to the corresponding activities on specimens in row B (middle of the holder). Each difference divided by average activity on the four specimens of each material type for normalization.

Judging by sight, there appears to be a distinction between the behavior of ^{51}Cr and that of the other three radionuclides. A statistically significant difference was observed between the means of the ^{51}Cr differences and those of ^{95}Nb, based on a t-test at the 95% confidence level. Mathematically, this statement is expressed as

$$\frac{a_{^{51}Cr,\,matX,rowB} - a_{^{51}Cr,matX,rowA}}{a_{^{51}Cr,matX,ave}} > \frac{a_{^{95}Nb,\,matX,rowB} - a_{^{95}Nb,matX,rowA}}{a_{^{95}Nb,matX,ave}}$$

There was no statistically significant difference in the means for ^{95}Zr or ^{181}Hf relative to those of ^{95}Nb. In mathematical terms, using ^{95}Zr as an example,

$$\frac{a_{^{95}Nb,\,matX,rowB} - a_{^{95}Nb,matX,rowA}}{a_{^{95}Nb,matX,ave}} \sim \frac{a_{^{95}Zr,\,matX,rowB} - a_{^{95}Zr,matX,rowA}}{a_{^{95}Zr,matX,ave}}$$

This comparison of means shows that the location dependence for ^{51}Cr activity is significantly different from the location dependence for ^{95}Nb, ^{95}Zr and ^{181}Hf, and that the ^{51}Cr activities were lower in the specimens that were farther from the flux.

Discussion

The activity of radionuclides on test specimens has been categorized in terms of the effect of corrosion rate of the specimen material. Some of the predominant species (^{60}Co, ^{58}Co, ^{59}Fe, ^{54}Mn and ^{65}Zn) show higher activity on the faster corroding alloys. This is consistent with the mechanism of incorporation into a growing film, and some of these species are known to be incorporated into corrosion films from other studies [2,9]. The higher activity in the faster corroding alloy is noted with the radionuclides of high activity (*e.g.*, ^{60}Co) as well as those with low activity (*e.g.*, ^{54}Mn). In all cases, the specimen-to-specimen variability is relatively low.

The activity of other radionuclides is insensitive to the corrosion rate of the test specimen material. The predominant species which demonstrate this behavior are ^{51}Cr, ^{95}Nb, ^{95}Zr and ^{181}Hf, but among these, the behavior of ^{51}Cr is unique. While each of these species shows higher specimen-to-specimen variability than do the species which are incorporated into corrosion films, only ^{51}Cr showed sensitivity to the specimen location. The lower activity on the samples farther from the flux had lower activity from those in the middle of the sample holder. This is consistent with the direct activation mechanism.

The sources of the other species are believed to be specimens from other types of tests which were going on in the same loop of the ATR at the same time. Records indicate that sources of Zr and Hf were present in other holders in the same loop, and ^{95}Nb comes from the activation of Zr. These species may have been transported as particulates which deposited non-specifically on the test specimens. This mechanism is less likely to be responsible for the observed activities of ^{51}Cr since Cr typically remains on substrate surfaces where it is first oxidized, as chromite (MCr_2O_4), and is not readily transported as a soluble cation or as a particulate in high-temperature water.

Conclusions

A review of historic activity pick-up data has revealed that the activities of several radionuclides increase with the corrosion rate of the material. This is evidence of activity pick-up by incorporation into a growing corrosion film, as these species are known to incorporate into corrosion films. The main species which showed this behavior are ^{60}Co, ^{58}Co, ^{59}Fe, ^{54}Mn and ^{65}Zn.

Of the species not incorporated into corrosion films, most of these must have come from external sources. Since they were unaffected by the material type, the postulated mechanism is non-specific adsorption. For example, the activity of ^{95}Nb was unaffected while the amount of Nb in the alloy changed by more than an order of magnitude. The main species which showed this behavior are ^{95}Nb, ^{95}Zr and ^{181}Hf.

Unlike the other species unaffected by corrosion rate, ^{51}Cr showed high location dependence. It is postulated that the mechanism of activity pick-up is direct activation. With this mechanism, the activity depends strongly on the location in the flux field and on the amount of the relevant species in the alloy. The effect of composition was not discernable, however, as all materials tested had between 17 and 25 wt% Cr, which is a relatively narrow range.

References

1. EPRI Report 1013509: "EPRI Alpha Monitoring Guidelines for Operating Nuclear Power Stations, Final Report," November 2006
2. EPRI Report 10025307: "Impact of PWR Coolant Radiocobalt Concentration on Shutdown Dose Rates, Interim Report," December 2012
3. J. Robertson, Corros. Sci. 32 (1991) 443
4. D. H. Lister, R. D. Davidson and E. McAlpine, Corros. Sci. 27 113 (1987)
5. S. E. Ziemniak, M. Hanson, "Electropolishing Effects on Corrosion Behavior of 304 Stainless Steel in High Temperature, Hydrogenated Water," Corros. Sci. 50 (2008) 2465
6. N. W. Grimes, "On the Specific Heat of Compounds with Spinel Structure. I. The Ferrites," Proc. R. Soc. London A 338 (1974) 209-221
7. H. St. C. O'Neill, A. Navrotsky, "Cation Distributions and Thermodynamic Properties of Binary Spinel Solid Solutions," Am. Mineral. 69 (1984) 733-753
8. S. E. Ziemniak, L. M. Anovitz, R. A. Castelli and W. D. Porter, "Thermodynamics of Cr_2O_3, $FeCr_2O_4$, $ZnCr_2O_4$ and $CoCr_2O_4$," J. Chem. Thermodynamics 39 (2007) 1474-1492
9. M. Juergensen, D. Sommer and B. Stellwag, "Zinc Injection for Further Reduction of Radiation Fields in German PWR Plants: A Status Report," Proceedings of the International Conference of Water Chemistry of Nuclear Reactor Systems 8," British Nuclear Energy Society, London, page 34 (2000)
10. S. E. Ziemniak, M. Hanson, "Corrosion Behavior of NiCrFe Alloy 600 in High Temperature, Hydrogenated Water," Corros. Sci. 48 (2008) 498-521

453595 Engineering Challenges in the Development of Actinide/Lanthanide Separation Processes to Support Advanced Nuclear Fuel Reprocessing

Abstract:

The separation of Am from the lanthanides and curium is a key step in proposed advanced fuel cycle scenarios. The partitioning and transmutation of Am is desirable to minimize the long-term heat load of material interred in a future high-level waste repository. Given only subtle chemistry differences within and between the ions of the trivalent actinide and lanthanide series, this separation is challenging; however, higher oxidation states of americium can be prepared and separated via solvent extraction. One such method utilizes sodium bismuthate oxidation and solvent extraction with diamylamylphosphonate (DAAP). Additional oxidants, including silver catalyzed ozone and copper(III) periodate have been investigated along with alternative extractants. Although testing of some of these processes in the laboratory has been successful, numerous chemical engineering challenges arise when developing the process flowsheets and separation equipment for process operations. Issues include solid sodium bismuthate in the aqueous feed solution, corrosivity of Ce(IV), kinetics issues arising from instability of Am(VI), and development of selective stripping methods for the Am and Ce. A "hot" engineering-scale centrifugal contactor test bed has been constructed and tested as well a laboratory-scale test bed, utilizing 3-D printing technologies. Recent research aimed at resolving the engineering challenges of implementation of a separation process based on higher oxidation states of Am will be presented.

471307 Extraction Kinetics and Hydrodynamics in Miniature Annular Centrifugal Contactors

Abstract:

Annular centrifugal contactors have become attractive equipment in the recycling process of spent nuclear fuel for their high efficiency, throughput, and rapid operational properties. Various sizes of contactors ranging from laboratory scale to commercial scale have been manufactured for different needs and process scales, possibly allowing us to study the scaling effects on the extraction efficiency. The Robatel BXP012 is a lab scale French made contactor with a 12mm inner rotor diameter. In this study, the hydrodynamics of the miniature contactor is studiued including residence time of both the aqueous and the organic phases by using residence time distribution analysis. The measured values proved not to reflect the residence time of the phases inside the mixing zone possibly due to the relatively large collection rings of the contactor. The effect of rotation speed, phase continuity, and flow rate on the extraction efficiency have also been investigated in this study. Comparison with the larger CINC contactor (50 mm rotor) is made and the results are discussed.

469066 Lab-Scale Tests of Codecontamination Processing

Abstract:

Codecontamination is a solvent extraction process that uses tributyl phosphate (TBP) to co-extract U, Pu, and Np from spent fuel dissolved in nitric acid, followed by selective stripping to recover U/Pu/Np in a single product stream, and the remaining U in a second effluent. For successful separation, the process requires predictable control of the behavior of the key species involved in the separations. In terms of flowsheet design, the major factors include controlling the U:TRU ratio in the U/Pu/Np product, directing the majority of the Np to the U/Pu/Np product, minimizing the extraction of fission products (specifically Mo, Ru, and Zr), and controlling Tc behavior. To that end, lab-scale flowsheet testing has been divided into three components: (1) demonstrating that extraction of the key fission products, Ru, Mo and Zr is controlled as predicted by the Argonne Model for Universal Solvent Extraction (AMUSE) code, (2) characterizing the extraction of Np as a function of redox chemistry, and (3) demonstrating the partial stripping of U in the second segment of the process. Multistage tests of the first and second segments of the Codecontamination flowsheet were run with a lab-scale 2-cm contactor to validate the design criteria used to develop the flowsheet. A number of small-scale tests were also conducted using a microfluidic extraction device. By varying flow rates and capillary lengths, mass transfer and kinetics data can be elucidated for U and Np in the TBP/alkane-nitric acid system. The behavior of U in the extraction is well-understood, which allows us to characterize the microfluidics device. Since Np demonstrates complicated redox chemistry that is difficult to regulate at large scales in this process, the microfluidic device was used to elucidate the Np extraction kinetics for different processing conditions.

471229 In-Line Near Real Time Monitoring of Fluid Streams in Separation Processes for Used Nuclear Fuel

Abstract:

Applying spectroscopic tools for monitoring chemical processes has been utilized in various industries owing to its rapid and non-destructively analysis for detecting chemical components and physical characteristic in a process stream. The general complexity of separation processes for used nuclear fuel, e.g., chemical speciation, temperature variations, and prominent process security and safety concerns, require a well-calibrated monitoring system to provide precise information of the process streams at real time without interference. Multivariate analysis accompanied with spectral measurements is a powerful statistic tool that can be used to monitor this complex chemical system. In this work, chemometric models that respond to the chemical components in the samples were calibrated and validated to establish an inline near real time monitoring system. The models show good prediction accuracy using partial least square regression analysis on the spectral data obtained from NIR and UV/Vis spectroscopies. The models were tested on an extraction process using a single stage centrifugal contactor in our laboratory to determine the performance of an inline near real time monitoring system for a solvent extraction process representative of used nuclear fuel separation processes. We present here the results from the online, near real-time monitoring of a simplified PUREX process using off-the-shelf components.

467241 Adsorption Process Modeling for the Recovery of Uranium from Seawater

Abstract:

One of the worldâ€™s rising global issues is how to meet the necessary energy demands of tomorrow without destroying the resources and environment of today. This challenge is being met by research into solar and wind energy, biofuels, and nuclear power. To facilitate the use of nuclear power in place of more traditional fossil-energy sources requires the utilization of the ocean as a new source of uranium. Adsorption technology can be employed to farm uranium from seawater in order to provide a new, untapped source of fuel for nuclear power. The ocean has the potential to provide a vast amount of uranium to add to the nuclear fuel cycle, as it contains nearly 1000 times more uranium than conventional mines. However, the average concentration of that resource is roughly 3.3 ppb, which presents a technological challenge. Recently, significant effort has been focused on the utilization of amidoxime ligands for selective removal of uranium from seawater, but there are still many other environmental factors such as pH, temperature, competing ions, and current flows that further complicate the design of a capture system. The realization of the ocean uranium recovery system will require an advanced understanding of the physicochemical processes involved, as well as a set of engineering modeling tools to aid in the process design. An advanced modeling framework coupled with detailed, high-resolution adsorption process models is being developed in order to give engineers a reliable design tool for the implementation of a seawater-capture system. These models are based on state-of-the-art linear and non-linear solver routines and are being designed specifically for generality and flexibility. Building of the framework in this manner allows us to branch off and develop more specific models for a particular system of interest, while also allowing the basic framework to be applicable to a wider variety of adsorption systems.

462729 First-Principles Based Computational Study on Nucleation and Growth Mechanisms of U on Mo(110) Surface Solvated in an Eutectic Licl–KCl Molten Salt

Abstract:

First-principles based computational study on nucleation and growth mechanisms of U on Mo(110) surface solvated in an eutectic LiCl–KCl molten salt

Choah Kwon, Byungchan Han*

Department of Chemical and Biomolecular Engineering, Yonsei University, Seoul 03722, Republic of Korea

We utilize first principles density functional theory (DFT) calculations and ab-initio molecular dynamic (AIMD) simulations to identify underlying mechanisms elucidating the initial stage of electrocrystallization process of U on Mo(110) surface in a eutectic LiCl–KCl molten salt at T = 773 K. Our results clearly unveil surprisingly different principles on the nucleation of U in the media from that under vacuum conditions. U nanoclusters exposed to vacuum completely collapse into flat atomic layers on Mo(110) surface similar to an electrodeposition process. On the other hand, Cl ions in eutectic molten salt thermodynamically drive crystallite formation consisting of UCl_n (n=3–6) through agglomeration of U atoms. Those crystallite gradually grows into bigger nuclei by adsorbing on Mo(110) surface. We propose that those behaviors are understandable only with revised conventional theories and that atomic level interactions among U, LiCl–KCl molten salt and Mo(110) surface play a key role to describe the atomic-scale dendrite formation of U in the electrorefining process. Our study can be one of the basic steps to design efficient electrorefining systems by identifying the fundamental cause of the experimentally observed uranium nucleation

phenomena.

468207 How Department of Energy - Environmental Management Stores, Treats and Disposes of Radiological and Mixed Hazardous Wastes at the Savannah River Site

Abstract:

The authors will present a paper/discussion of how Environmental Management stores, treats and disposes of radiological and mixed hazardous wastes at the Savannah River Site. It will include an overview of waste streams and treatment processes, including significant challenges.

SRNL-STI-2016-00554

DISCLAIMER

Printed in the United States of America

Prepared for
U.S. Department of Energy

IMPLEMENTATION OF FLOWSHEET CHANGE TO MINIMIZE HYDROGEN AND AMMONIA GENERATION DURING CHEMICAL PROCESSING OF HIGH LEVEL WASTE IN THE DEFENSE WASTE PROCESSING FACILITY

Dan P. Lambert, Wesley H. Woodham, Matthew S. Williams, J. David Newell, Savannah River National Laboratory, Aiken, SC, USA, Michelle C. Luther, Auburn University, Auburn, AL, USA, Clayton H. Brandenburg, University of South Carolina, Columbia, SC, USA

Abstract

Testing was completed to develop a chemical processing flowsheet for the Defense Waste Processing Facility (DWPF), designed to vitrify and stabilize high level radioactive waste. DWPF processing uses a reducing acid (formic acid) and an oxidizing acid (nitric acid) to rheologically thin the slurry and complete the necessary acid base and reduction reactions (primarily mercury and manganese). Formic acid reduces mercuric oxide to elemental mercury, allowing the mercury to be removed during the boiling phase of processing through steam stripping. In runs with active catalysts, formic acid can decompose to hydrogen and nitrate can be reduced to ammonia, both flammable gases, due to rhodium and ruthenium catalysis. Replacement of formic acid with glycolic acid eliminates the generation of rhodium- and ruthenium-catalyzed hydrogen and ammonia. In addition, mercury reduction is still effective with glycolic acid. Hydrogen, ammonia and mercury are discussed in the body of the report.

Ten abbreviated tests were completed to develop the operating window for implementation of the flowsheet and determine the impact of changes in acid stoichiometry and the blend of nitric and glycolic acid as it impacts various processing variables over a wide processing region. Three full-length 4-L lab-scale simulations demonstrated the viability of the flowsheet under planned operating conditions. The flowsheet is planned for implementation in early 2017.

Hydrogen

The offgas was monitored by a gas chromatograph and mass spectrometer for hydrogen. Hydrogen is generated in nitric-glycolic acid processing; however its concentration is almost always below the 0.006 volume % quantification limit of the gas chromatograph. In runs where hydrogen was detected but was below the quantification limit, the results will be reported. In runs where no hydrogen was detected, the results will be reported as less than the quantification limit. Two of the runs had thick rheology, fouling, and excessive heating rod temperatures. In these two runs where significant hydrogen was quantified, a note will be added saying the hydrogen data was collected at rod temperatures in excess of DWPF steam coils. Peak hydrogen generation data are summarized in

Table 1. The hydrogen concentration profile is shown in Figure 1.

Table 1 Peak Hydrogen Concentration and Generation Rates

Experiment #	Acid Stoichiometry	Percent Reducing Acid	H_2 Peak mmol/min	H_2 Peak vol %	H_2 Total mmol
51	83.7	58.36	<0.0018	<0.006	N/A
52*	116	54.1	0.0048	0.032	0.6
53	83.6	62.62	0.0005	0.003	0.4
54*	117	58.36	0.0410	0.227	1.5
55	100	58.36	0.0004	0.003	0.3
55A	100	52.34	0.0004	0.002	0.3
56	100	64.39	0.0004	0.003	0.2
57	99.9	62.62	0.0005	0.003	0.5
58	76.9	62.62	<0.0018	<0.006	N/A
59	123	62.62	<0.0018	<0.006	N/A
60	100	62.62	<0.0018	<0.006	N/A
61	100	54.1	<0.0018	<0.006	N/A
62	97.4	58.36	0.0007	0.004	0.9

* Hydrogen data generated while heating rod temperatures exceeded 160 °C.

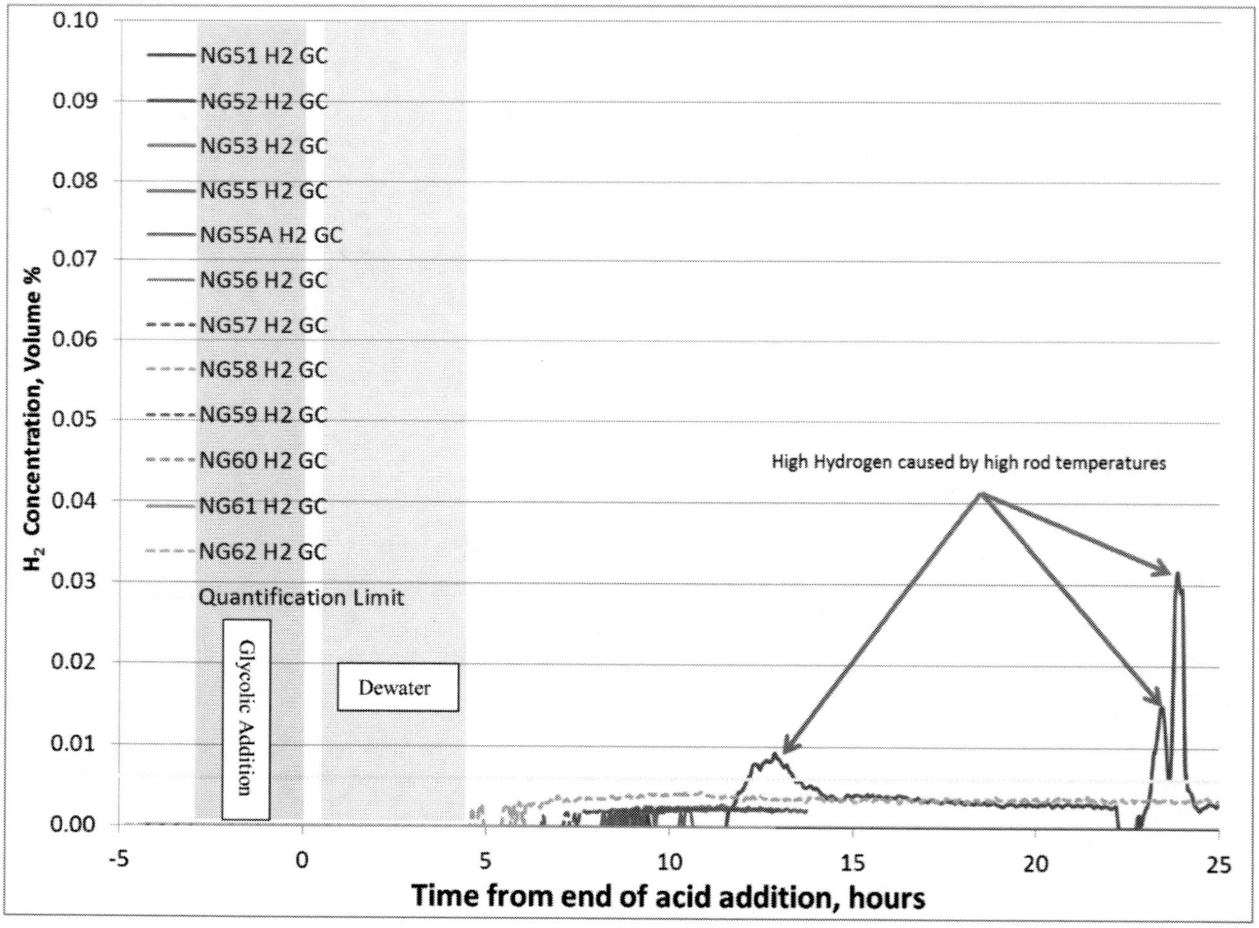

Figure 1 Hydrogen Concentration All Runs, Volume %

Ammonia

Ammonia scrubbers are operated to scrub ammonia from the offgas use the acidic condensate to prevent the formation of ammonium nitrate (an explosive hazard) solids in the offgas piping. Ammonia can be generated during chemical processing and is released to the offgas when the ammonia/ammonium equilibrium favors ammonia (slurry pH >7). If the slurry pH is acidic, the ammonia is retained in the slurry as ammonium. The offgas was monitored by a Fourier transform infrared spectrometer (FTIR), and mass spectrometer.

In addition to the offgas, other sources for ammonia are the products and the ammonia scrubber solution. The concentrations of ammonium in the scrubber solutions and the concentrations of ammonium in the products are summarized in Table 2. Note that no ammonium was detected in the ammonia scrubber solutions. This indicates that if ammonium is being produced, the ammonia/ammonium equilibrium is preventing the release of ammonia to the offgas. Note also that a low concentration of ammonium was detected in runs 61 and 62. Since the slurry pH was acidic, no detectable ammonium was adsorbed by the ammonia scrubber. It is also interesting that the runs that lasted longer, runs 61 and 62, produced more ammonium.

Table 2 Ammonium in Ammonia Scrubber Solution and Products, mg/L

	51	52	60	61	62
Post Acid Scrub Solution	NA	NA	<5	<5	<5
Post Processing Scrub Solution	NA	NA	<5	<5	<5
Post Processing Slurry	NA	NA	<5	6.39	11.5
Post Processing Slurry pH	7.45	4.36	5.00	4.93	5.93

The result is that little ammonia is produced during simulant testing and the ammonia is retained as ammonium by the slurry since it is acidic. Thus, the main function of the scrubbers under the glycolic-nitric flowsheet is to remove NO2 gas as nitric acid, as well as scrub other volatile species (such as mercury) from offgas streams. Because the pH of the scrubber need not be regulated to optimize ammonia removal under the nitric—glycolic flowsheet, the pH of the scrub solution may be changed to improve scrubbing efficiencies of the alternate species."

Mercury

DWPF is designed to be the purge point for mercury in high level waste processing. For this to succeed the mercury needs to be reduced, steam stripped, coalesced, and collected in the mercury decanter or condensate tank. In addition, once the elemental mercury is collected, it can't be dissolved or reoxidized or it will be reprocessed at a later date as the condensate is processed. The recovery of mercury in the mercury decanter has been extremely poor in DWPF. During this testing, mercury stripping and recovery were quantified for each run. In an attempt to increase the mercury recovery, all runs maintained an agitator speed of 700 revolutions per minute (rpm). In addition, all runs except 61 used the maximum boilup rate of 5,000 lb/hr.

The mercury decanter is not an ideal decanter (slow liquid flows yielding sufficient residence times for the mercury to coalesce). Instead, it is a busy vessel where NO_x is scrubbed, mercury is dissolved, other reactions occur, and water is continually flowing (dropping) during conflux. A photo of the mercury decanter used in this testing during the evaporation phase of processing is shown in Figure 2.

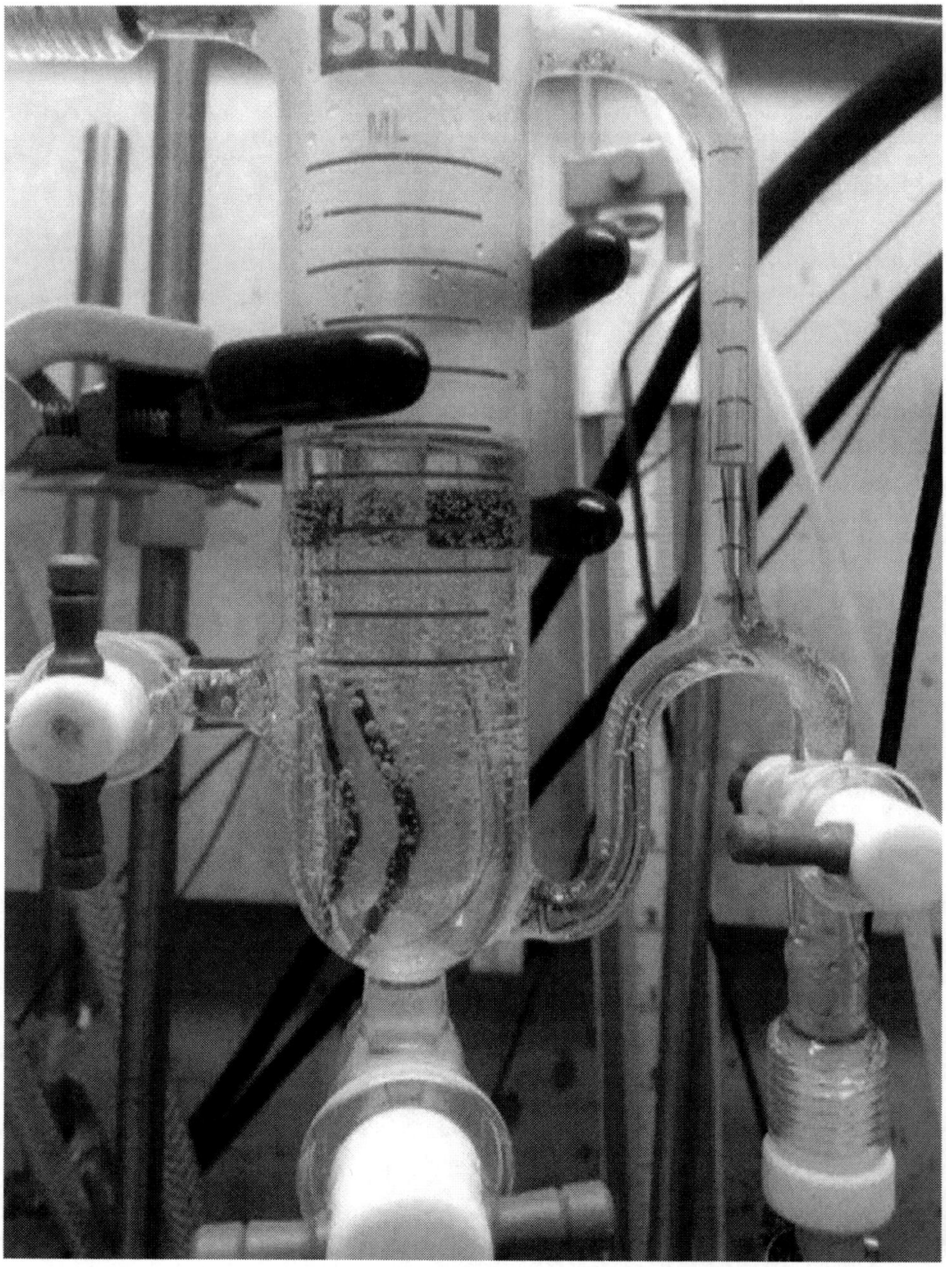

Figure 2 Mercury Decanter during Dewater

In each of the thirteen experiments, approximately 14.7 g of mercuric oxide (equivalent to 13.6 g of elemental mercury) was added to target a mercury concentration of 2.48 wt% total solids basis in the sludge simulant. One goal of the processing is to reduce the mercury loading by steam stripping the mercury to less than 0.45 wt % total solids basis in the product. In order to reach this product mercury target, steam stripping (time at boiling or conflux) is planned for approximately 36 hours at a scaled boilup rate of 5,000 lb/hr condensate or 72 hours at a scaled boilup rate of 2,500 lb/hr, assuming it takes 750 g of steam to strip 1 g elemental Hg.

Slurry samples were pulled throughout the testing. Mercury is only stripped during boiling. During evaporation segment of processing, the stripping rate is constant but the mercury concentration increases due to evaporation of water. So although mercury is being stripped (between 0 and 5 hours for 5000 lb/hr scaled steam flow or between 0 and 10 hours for 2500 lb/hr scaled steam flow), the mercury concentration increases from about 3300 mg/kg to 4200 mg/kg. Once evaporation is complete, the condensate is returned to the reactor and the mercury concentration is expected to decrease linearly until boiling is complete (targeting a final mercury concentration of 1125 mg/kg or 0.45 wt %). A projected mercury stripping line for the maximum boilup runs (60 and 62, purple line for 5000 lb/hr scaled steam flow) and for the prototypic boilup run (61, orange line for 2500 lb/hr scaled steam flow) showing the mercury concentration versus time at boiling for Runs 60, 61 and 62 is summarized in Figure 3.

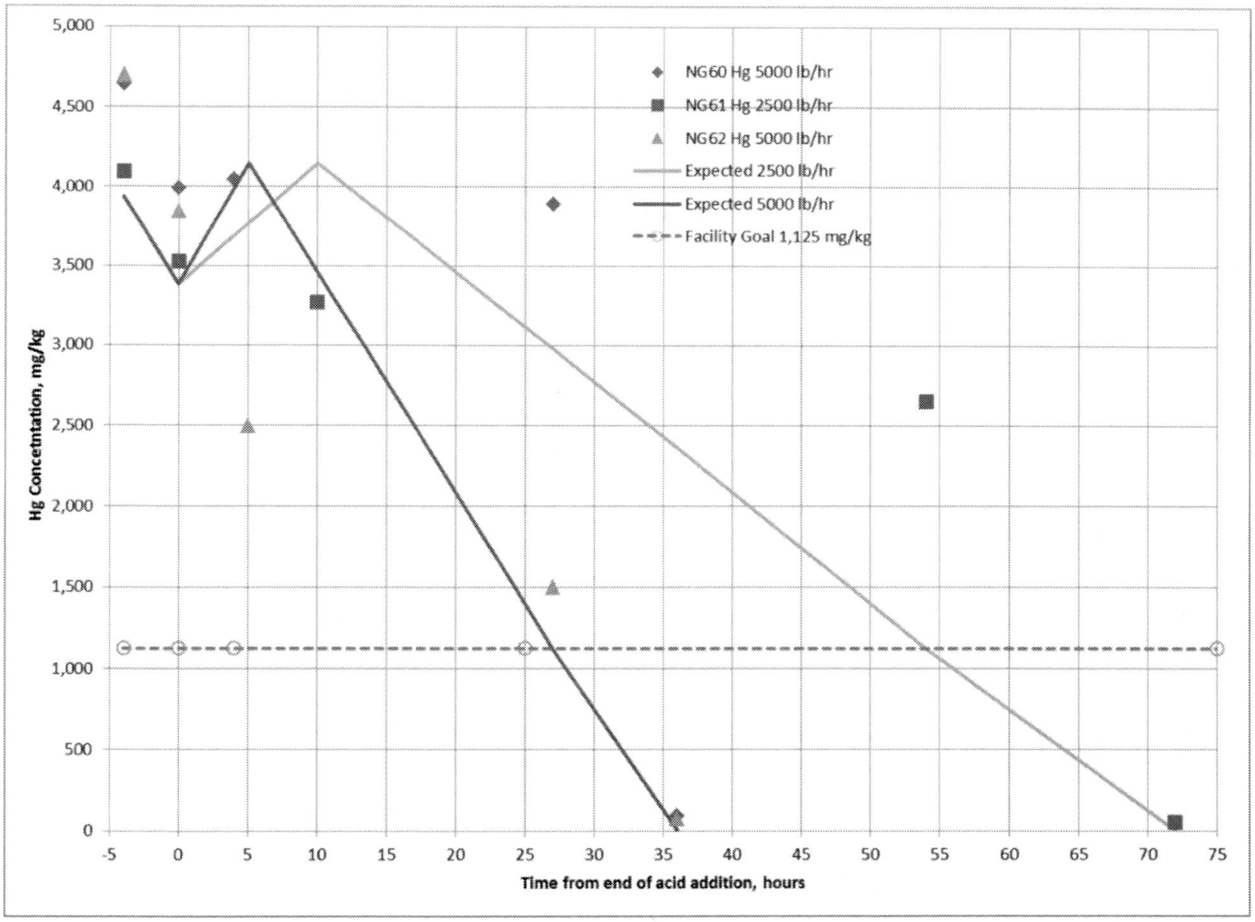

Figure 3 Mercury Concentration during Processing

At the completion of each run, the mercury decanter contents were drained to a sample bottle. The aqueous liquid in the sample bottle was transferred to a second sample bottle, leaving the mercury (and a small amount of water) in the first sample bottle, which was placed in a desiccator for at least a week. The dried mercury was weighed. The mercury added, the mercury collected from the decanter, the % of mercury recovered in the decanter, the measured mercury in the product, and the calculated mercury from each run is summarized in Table 3. The mercury in the product was calculated assuming any mercury not collected in the mercury decanter is still in the product. The runs are ordered top to bottom from the lowest acid stoichiometry to the highest acid stoichiometry.

Table 3 Mercury Added, recovered, and calculated mercury concentration in Product

Run	Boilup Rate, lb/hr DWPF Scale	Hg Added, g	% Recovered in mercury decanter	Measured Hg in Product		Calculated Hg In Product, mg/kg
				mg/kg*	wt % TS	
58	5000	13.58	53.9	64	0.021	4,380
51	5000	13.58	30.7	170	0.055	2,719
53	5000	13.58	39.6	270	0.090	3,724
55	5000	13.58	80.8	213	0.062	3,122
55A	5000	13.53	57.8	82	0.041	1,305
56	5000	13.58	81.0	432	0.140	1,682
57	5000	13.59	60.9	135	0.044	1,161
60	5000	13.57	73.9	269	0.105	2,430
61	2500	13.58	22.8	274	0.102	2,942
62	5000	13.58	29.2	108	0.042	632
52	5000	13.58	50.3	1,610	0.610	1,358
54	5000	13.58	43.5	NA	NA	4,127
59	5000	13.58	82.2	195	0.104	3,153

* Aqua Regia digestion of entire sample followed by ICP-AES analysis for Hg

The mercury recovery was much higher than expected (typical mercury recovery is ~30%). In these runs, the mercury recovery was as high as 81%. What is particularly noteworthy is the difference in mercury collection between runs 60 and 61, with the only difference being the boilup rate. Run 60 had 3x higher mercury recovery than 61. This may demonstrate the importance of high steam flow on mercury recovery. It is also expected that agitation speed is key to improved mercury recovery, although this was not varied in these runs. Good mixing and high boilup rates should work together to disperse the mercury more uniformly and should lead to better steam stripping. One other surprise was the lower mercury recovery in 62. Little testing has been done with the two additional waste streams that were added during run 62. More testing is recommended to determine the repeatability of the lowered mercury recovery under this type of processing.

Elemental mercury is the assumed form of mercury collected in the mercury decanter. If other forms or other impurities are present, they could change the reported mercury recovery. The mercury from runs 51 and 52 were completely digested and analyzed by Inductively Coupled Plasma-Atomic Emission Spectroscopy (ICP-AES). The results are reported in Table 4. Si was the only metal reported above the 10 mg/kg level. The sample collected from the mercury decanter was measured and reported as 1.00E6 mg/kg Hg and was therefore determined to be primarily elemental mercury.

Table 4 Impurities in Mercury, mg/kg

	BLANK		51 Mercury		52 Mercury	
Element	**Result**	**Uncertainty**	**Result**	**Uncertainty**	**Result**	**Uncertainty**
Ag	< 0.735	(N/A %RSD)	3.77	(13.8 %RSD)	< 0.726	(N/A %RSD)
Al	**< 1.52**	**(N/A %RSD)**	**1.44**	**(110 %RSD)**	**3.84**	**(13.6 %RSD)**
B	< 0.53	(N/A %RSD)	< 0.483	(N/A %RSD)	< 0.524	(N/A %RSD)
Ba	< 0.083	(N/A %RSD)	< 0.076	(N/A %RSD)	< 0.082	(N/A %RSD)
Ca	**< 0.119**	**(N/A %RSD)**	**0.336**	**(10 %RSD)**	**0.387**	**(13 %RSD)**
Cd	< 0.144	(N/A %RSD)	< 0.131	(N/A %RSD)	< 0.142	(N/A %RSD)
Ce	< 2.29	(N/A %RSD)	< 2.09	(N/A %RSD)	< 2.26	(N/A %RSD)
Co	< 0.19	(N/A %RSD)	< 0.173	(N/A %RSD)	< 0.187	(N/A %RSD)
Cr	< 0.083	(N/A %RSD)	< 0.076	(N/A %RSD)	< 0.082	(N/A %RSD)
Cu	10.9	(11 %RSD)	5.82	(14.3 %RSD)	0.761	(115 %RSD)
Fe	**< 0.096**	**(N/A %RSD)**	**1.85**	**(10.3 %RSD)**	**5.53**	**(10.3 %RSD)**
K	< 12.1	(N/A %RSD)	< 11	(N/A %RSD)	< 11.9	(N/A %RSD)
La	< 0.332	(N/A %RSD)	< 0.303	(N/A %RSD)	< 0.328	(N/A %RSD)
Li	< 0.16	(N/A %RSD)	< 0.146	(N/A %RSD)	< 0.158	(N/A %RSD)
Mg	< 0.051	(N/A %RSD)	< 0.046	(N/A %RSD)	< 0.05	(N/A %RSD)
Mn	< 0.206	(N/A %RSD)	< 0.204	(N/A %RSD)	< 0.204	(N/A %RSD)
Mo	< 0.663	(N/A %RSD)	< 0.604	(N/A %RSD)	< 0.655	(N/A %RSD)
Na	< 1.18	(N/A %RSD)	< 1.08	(N/A %RSD)	< 1.17	(N/A %RSD)
Nb	< 0.475	(N/A %RSD)	< 0.433	(N/A %RSD)	< 0.469	(N/A %RSD)
Nd	< 0.804	(N/A %RSD)	< 0.734	(N/A %RSD)	< 0.795	(N/A %RSD)
Ni	< 0.156	(N/A %RSD)	< 0.142	(N/A %RSD)	< 0.154	(N/A %RSD)
P	< 0.896	(N/A %RSD)	< 0.817	(N/A %RSD)	< 0.885	(N/A %RSD)
Pb	< 0.373	(N/A %RSD)	< 10	(N/A %RSD)	< 10	(N/A %RSD)
Re	< 0.51	(N/A %RSD)	< 0.465	(N/A %RSD)	< 0.504	(N/A %RSD)
S	< 0.677	(N/A %RSD)	< 0.618	(N/A %RSD)	< 0.669	(N/A %RSD)
Si	**6.2**	**(26.5 %RSD)**	**25.1**	**(10.3 %RSD)**	**28**	**(10.5 %RSD)**
Sn	< 2.72	(N/A %RSD)	< 2.48	(N/A %RSD)	< 2.69	(N/A %RSD)
Sr	< 4.9	(N/A %RSD)	< 4.47	(N/A %RSD)	< 4.84	(N/A %RSD)
Ti	< 0.156	(N/A %RSD)	< 0.143	(N/A %RSD)	< 0.154	(N/A %RSD)
V	< 0.261	(N/A %RSD)	< 0.238	(N/A %RSD)	< 0.258	(N/A %RSD)
Zn	< 0.326	(N/A %RSD)	< 0.297	(N/A %RSD)	< 0.322	(N/A %RSD)
Zr	< 0.227	(N/A %RSD)	< 0.207	(N/A %RSD)	0.364	(23 %RSD)

In some of the later runs, the product slurry was carefully poured out in an attempt to recover any unstripped elemental mercury. In Run 61 (a run with poor mercury recovery), 6.22 g of elemental mercury was recovered from the product. This confirms the poor mercury recovery in this run. Contrast this to Run 55, a run with excellent mercury recovery where only 1.22 g of mercury was recovered from the product. This suggests that the rate limiting step for mercury removal is physical and not chemical. A mercury mass balance was completed for all the runs and the results are summarized in Table 5.

Table 5 Mercury Mass Balance

Run	Elemental Hg Added, g	Mercury collected - mercury decanter, g	Mercury Collected - Product, g	Mercury in Empty Kettle, g	Total Out, g	% Hg Balance
58	13.58	7.32	0.14	NA	7.46	54.9%
51	13.58	4.17	0.37	NA	4.54	33.4%
53	13.58	5.38	0.59	NA	5.97	44.0%
55	13.58	10.98	0.42	1.09*	12.50	92.0%
55A	13.53	7.82	0.28	NA	8.10	59.9%
56	13.58	11.00	0.96	0.87*	12.83	94.5%
57	13.59	8.28	0.29	NA	8.57	63.1%
60	13.57	10.03	0.70	NA	10.73	79.1%
61	13.58	3.10	0.70	6.22*	10.02	73.8%
62	13.58	3.97	0.33	NA	4.30	31.7%
52	13.58	6.83	4.00	NA	10.83	79.7%
54	13.58	5.91	NA	NA	5.91	43.5%
59	13.58	11.16	0.75	NA	11.91	87.7%

* In some of the runs the product contents were poured out slowly, leaving elemental Hg behind. This Hg was weighed and the mass included.

The impact of acid stoichiometry and percent reducing acid on process chemistry and rheology were two of the objectives of this testing. A graph showing the concentration of mercury in the products is included as Figure 4.The higher acid stoichiometry leads to higher mercury recovery. The two high acid runs with low Hg recovery were 52 and 54, both with very thick rheology which likely hindered steam stripping. Note that the highest acid stoichiometry run (59) had similar recovery to the mid acid runs but had to be processed at 20 wt % total solids target to eliminate the problems with high rheology. This suggests that processing the slurry at the optimum acid stoichiometry and total solids should lead to higher mercury recovery.

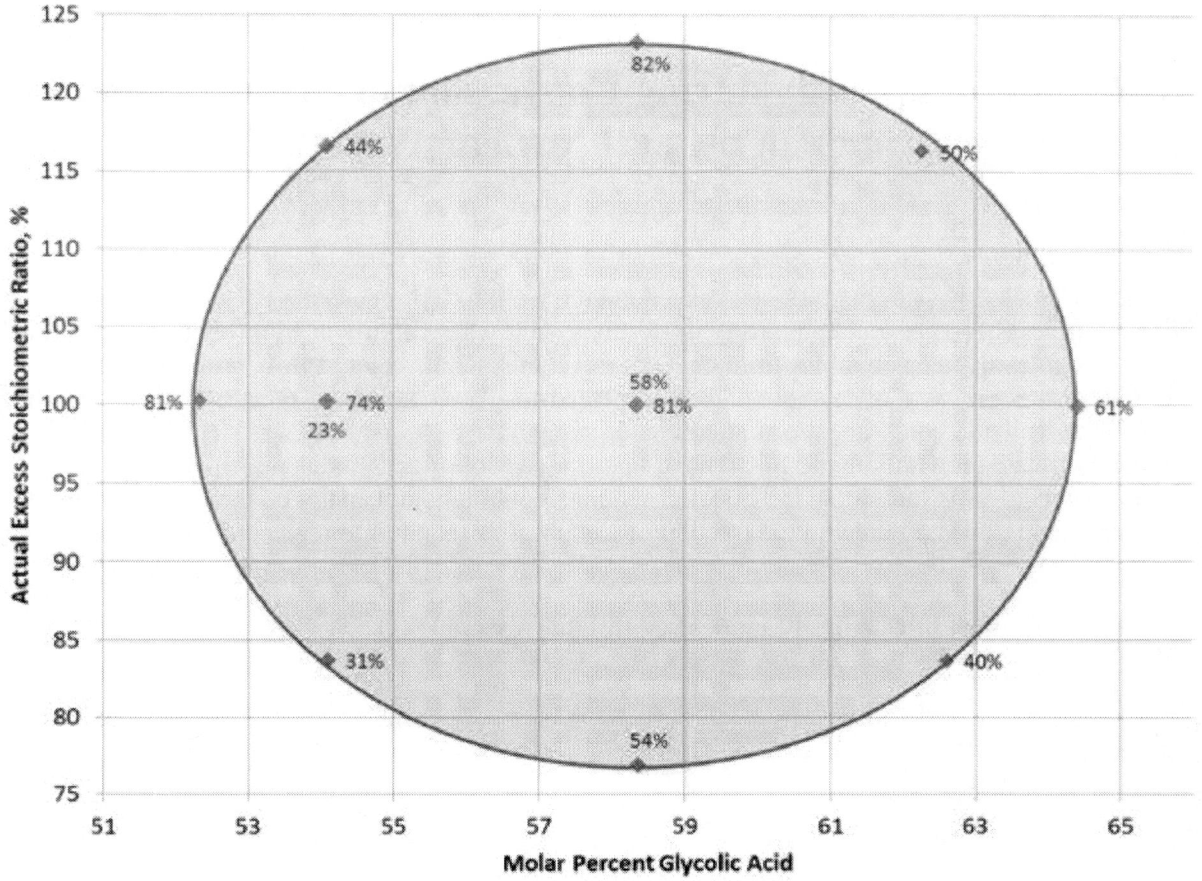

Figure 4 Percent Mercury Recovery as a Function of Acid Stoichiometry and Molar Percent Glycolic Acid

Conclusions

Testing was completed to develop a chemical process flowsheet for DWPF. Simulations were completed using simulants and radioactive waste. As has been demonstrated in over 100 simulations, the replacement of formic acid with glycolic acid virtually eliminates the largest flammability hazards, hydrogen and ammonia.

Implementation of the nitric-glycolic acid flowsheet in DWPF is recommended. This flowsheet not only eliminates the hydrogen and ammonia hazards but also will lead to shorter processing times, higher elemental mercury recovery.

464991 Melter Feed Viscosity during Conversion to Glass: Comparison Between Low-Activity and High-Level Nuclear Waste Feeds

Abstract:

During vitrification of nuclear wastes, a glass melter feed (a slurry mixture of nuclear wastes with glass-forming and glass-modifying additives) becomes a continuous glass-forming melt where undissolved refractory constituents are suspended together with evolved gas bubbles from complex reactions. Knowledge of flow properties of various reacting melter feeds is necessary to understand their unique feed-to-glass conversion processes occurring within a floating layer of melter called a cold-cap. We studied the viscosity of two low activity waste (LAW) melter feeds during heating and correlated it with volume fractions of undissolved solids and gas phase. In contrast to the high level waste (HLW) melter feed, it was found that the effect of compositional inhomogeneity on the LAW melter feed viscosity is relatively negligible and the volume fraction of gas phase is required to represent the viscosity of LAW melter feeds.

468332 Crossflow Filtration of Waste Simulants in Support of the Low Activity Waste Pretreatment System

Abstract:

Crossflow filtration (CFF) is a unit operation selected to perform separation of solids (expected to be < 0.8 wt% when fed) from the waste streams retrieved from the Hanford Tank Farms in the Low Activity Waste Pretreatment System (LAWPS) facility. Prediction of CFF performance for waste treatment (filter flux or flow rate) is complex as several factors need to be considered: solid particle size distribution and morphology, solution chemistry, operational parameters, and filter history. Historically, this has motivated experimental/empirical studies; however, the majority of data collected for crossflow filtration using actual waste or waste simulants have focused either on solids loadings greater than ~5 wt%, i.e. HLW streams, or CFF dewatering performance (where the solids loading changes with time). Experimental CFF work using actual waste or waste simulants with low solids (< 3 wt%) has been more limited. The primary concern with CFF of low solids slurries is the potential for more significant deep fouling, since there may not be enough solids present to form a protective cake on the filter/fluid interface. It is not well understood what parameters control the occurrence of deep fouling in these crossflow filtration systems and how filter performance is affected over longer periods of operation.

To address this gap in understanding, Pacific Northwest National Laboratoryâ€™s bench-scale crossflow filtration system was used to examine performance at low solids loading in waste simulants developed to be representative of expected LAWPS feed. The bench-scale system was also modified to test a backflush cleaning approach proposed for use in the LAWPS facility. Resulting data from these tests, as well as comparisons with historical crossflow filtration data, will be presented and discussed.

449301 Challenges and Effective Methods for Implementing Full ASME Nqa-1 2008, 2009 Requirements on the Biggest, First-of-a-Kind Nuclear Projects.

Abstract:

Large, complex projects often require implementation of ASME NQA-1 (NQA-1) requirements, but little or no direction is given to effective implementation including the graded approach to quality. These requirements can become overwhelming to process engineers in the design of new plants. Using experience from large, complex nuclear plants like the Waste Treatment Plant (WTP) in Hanford Washington and the Uranium Processing Facility (UPF) in Oak Ridge, TN, the authors present an evaluation of the challenges and interpretations in implementing various requirements of NQA-1. While each site is unique, there are significant similarities including that much of the existing historical data will not have been gathered as part of an NQA-1 testing program. Crediting current operations, historical data, and gathering new data for process design represent major challenges to process designs invoking the consensus standards of NQA-1.

In the United States, customer interpretations by regulatory agencies often differ. The Nuclear Regulatory Committee, Dept. of Energy Environmental Management, Dept of Energy-NNSA and the Defense Nuclear Facilities Safety Board (DNFSB) all have different interpretations of the graded approach and quality expectations to the processes defined within the standard.

This paper/presentation presents practical applications designed to overcome these challenges, what has worked, what has required rework, and best-in-class practices to avoid QA surprises toward the end of a project.

NEUP Nuclear Energy University Programs
U.S. Department of Energy

OAK RIDGE National Laboratory

Co-Adsorption of Iodine and Water by Reduced Silver-Exchanged Mordenite

Yue Nan,[1] **Austin P. Ladshaw,[2] Sotira Yiacoumi,[2] Costas Tsouris,[2,3] David W. DePaoli[4] and Lawrence L. Tavlarides[1]***

[1]Department of Biomedical and Chemical Engineering, Syracuse University
[2]School of Civil and Environmental Engineering, Georgia Institute of Technology
[3]Energy and Transportation Science Division, Oak Ridge National Laboratory
[4]Nuclear Security and Isotope Technology Division, Oak Ridge National Laboratory

16AIChE
Annual Meeting, San Francisco, CA
November 15, 2016

College of Engineering and Computer Science
SYRACUSE UNIVERSITY

COLLEGE OF ENGINEERING AND COMPUTER SCIENCE | SYRACUSE UNIVERSITY

THE TEAM

The off-gas sorption modeling and verification team:

- ❑ I_2 and HTO equilibrium and kinetic data acquisition and modeling
  - ❑ Yue Nan and Lawrence L Tavlarides (PI)
- ❑ Development of complex models for adsorption of HTO, I_2, Kr and Xe
  - ❑ Austin P Ladshaw, Sotira Yiacoumi (Co-PI) and Costas Tsouris (Co-PI)
- ❑ HTO and I_2 column experiments and data acquisition
  - ❑ David W DePaoli (Co-PI) and Costas Tsouris (Co-PI)
- ❑ CO2 absorption experiments and modeling
  - ❑ Jorge Gabitto (Co-PI)

NEUP Nuclear Energy University Programs
U.S. Department of Energy

This work is supported by the Office of Nuclear Energy, US DOE, under Nuclear Energy University Program.
Contract #: NE0008275.

COLLEGE OF ENGINEERING AND COMPUTER SCIENCE | SYRACUSE UNIVERSITY

OFF-GAS MODELING AND VERIFICATION

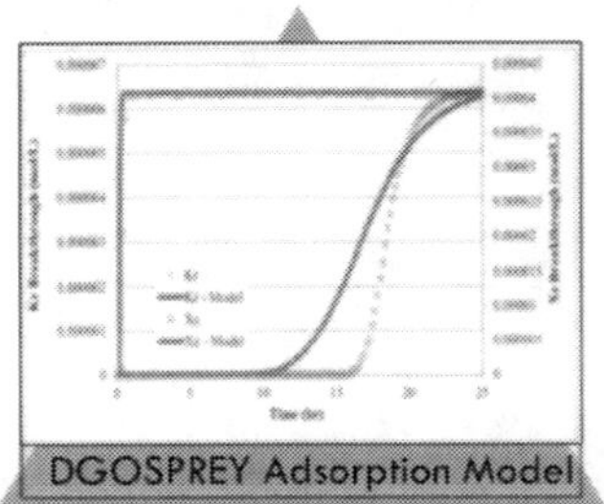

- ❑ Development of modeling tools for off-gas capture from nuclear fuel recycling facilities.
- ❑ Target species: I (I_2, CH_3I), ^3H (HTO), Kr, Xe, and C(CO).

I_2 & H_2O

Ag^0Z

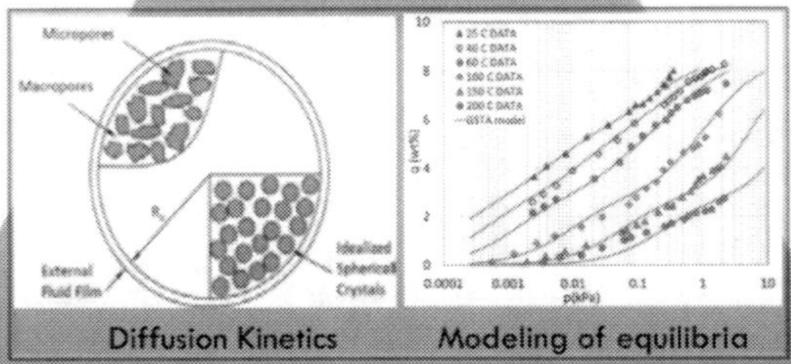

- ❑ Fundamental modeling of kinetics and equilibria
- ❑ Kinetic and isothermal parameters

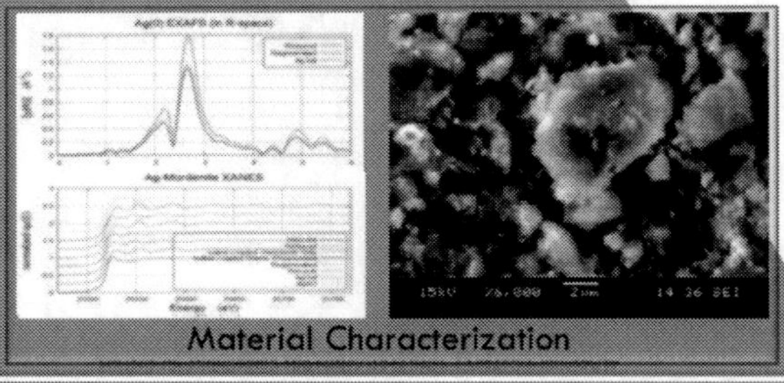

- ❑ Thin-bed experiments
- ❑ Characterization

ADSORBENT AND EXPERIMENTAL SYSTEM

- ❑ Ag^0Z (IONEX-Type Ag-900)
- ❑ Diameter: 1.8 mm
- ❑ Reduced at 400 °C, for 24 h
- ❑ Ag content: 11.9 wt.%
- ❑ Theoretical I_2 loading:14 wt.%

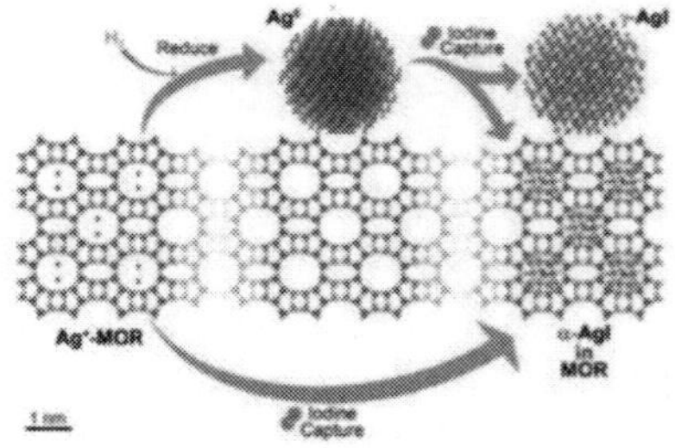

Iodine adsorption on Ag^0Z[1]

Screen tray with Ag^0Z loaded

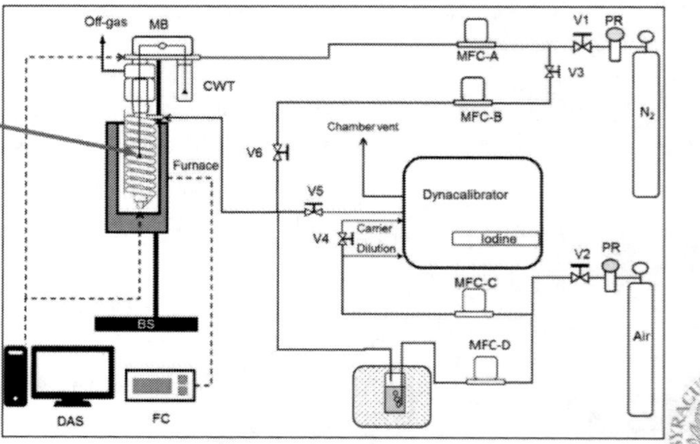

1.Chapman, K. et al.(2010). *JACS*, 132(26), 8897-8899.

IODINE UPTAKE CURVES

- Adsorption temperatures: 100, 150, 175 and 200 °C
- Gas stream dew point: -70 °C (dry air)

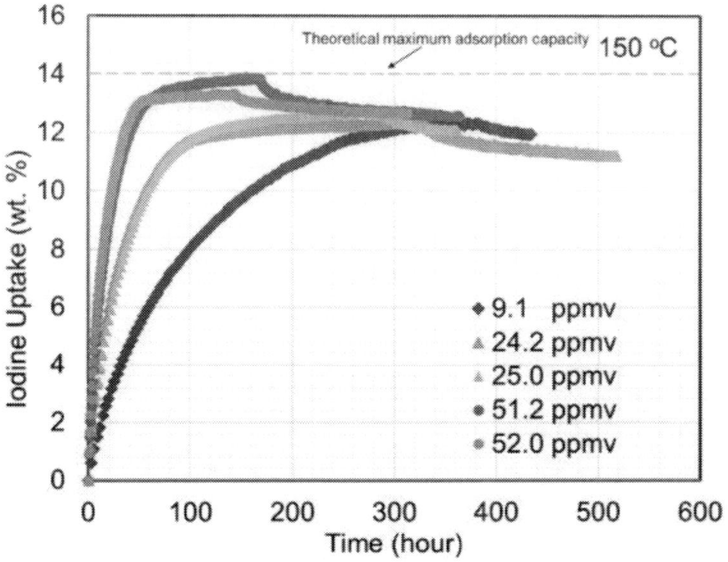

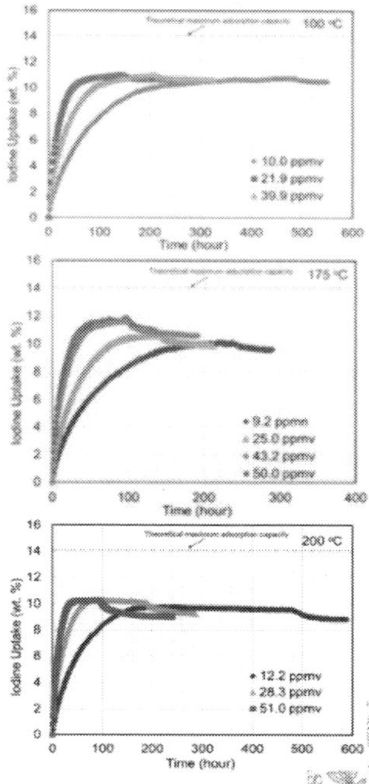

2. Nan, Y., et.al. (2016) *AIChE Journal.*

KINETIC MODELING OF I_2 ADSORPTION

Shrinking-Core (SC) Model:

$$t = \frac{q}{q_e}\tau_1 + \left(1 + 2\left(1 - \frac{q}{q_e}\right) - 3\left(1 - \frac{q}{q_e}\right)^{2/3}\right)\tau_2 + \left(1 - \left(1 - \frac{q}{q_e}\right)^{1/3}\right)\tau_3$$

$$\tau_1 = \frac{r_a \rho_B}{3bk_f C_A} \qquad \tau_2 = \frac{r_a^2 \rho_B}{6bD_p C_A} \qquad \tau_3 = \frac{r_a \rho_B}{bk_s C_A}$$

Gas-film mass transfer | Pore diffusion | Reaction

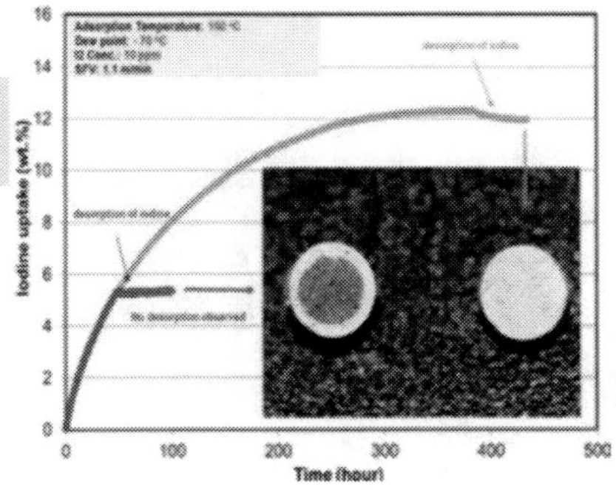

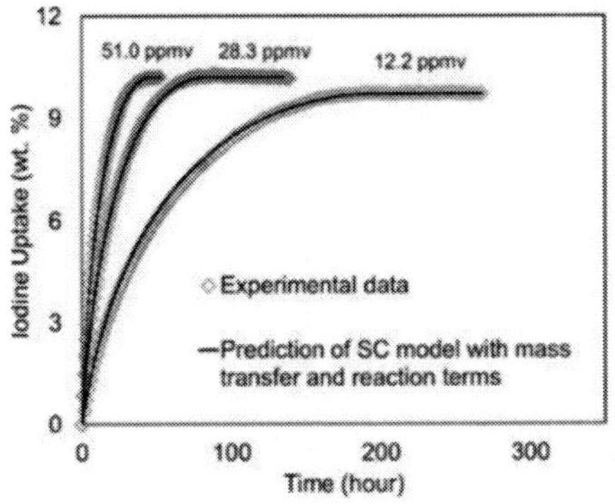

- k_f: 6.48 ~ 8.41 cm/s
- D_p: 0.00254 ~ 0.00416 cm^2/s
- k_s: 0.227 ~ 0.343 cm/s

3. Yue Nan et al, *AIChE*, 2015.

MODELING OF I_2 ISOTHERMS

GSTA Isotherm:

- ❑Multi-site model related to Heterogeneous Langmuir Model
- ❑Each site has the same maximum adsorption capacity
- ❑Known to be effective for water vapor isotherms in MS3A[4]

$$q = \frac{q_{\max}}{m} \frac{\sum_{n=1}^{m} n K_n p^n}{1 + \sum_{n=1}^{m} K_n p^n}$$

GSTA Isotherm *(Llano-Restrepo and Mosquera, 2009)*
Code optimizes for q_{max}, m, and each K_n simultaneously
When m=1, GSTA simplifies to Langmuir Model

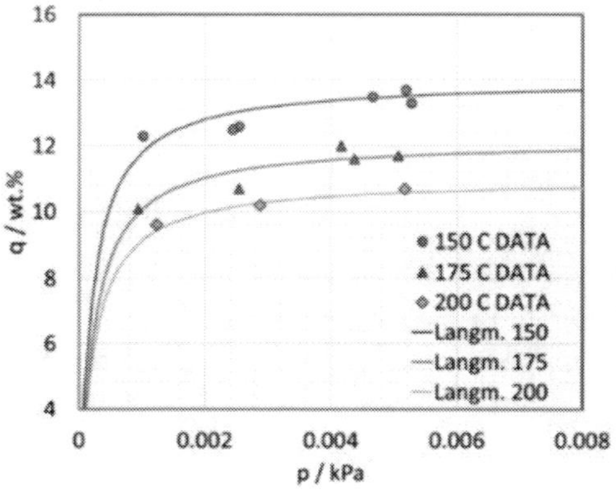

Curve-fittings of Langmuir model with iodine adsorption experimental data

4. Lin, R., et.al. (2014). I&ECR. 54, 10442-10448.

KINETICS OF H_2O ADSORPTION

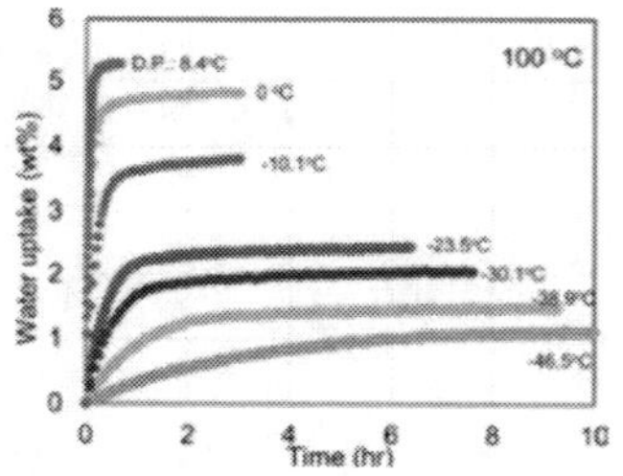

- ❑ Adsorption temperatures: 25 ~ 200 °C
- ❑ Dew points: -54 – 12.4 °C
- ❑ Maximum monolayer capacity: 8.56 wt.%

Shrinking-Core (SC) Model:

$$t = \frac{q}{q_e}\tau_1 + \left(1 + 2\left(1 - \frac{q}{q_e}\right) - 3\left(1 - \frac{q}{q_e}\right)^{2/3}\right)\tau_2$$

$$\tau_1 = \frac{r_a \rho_B}{3 k_f C_A} \qquad \tau_2 = \frac{r_a^2 \rho_B}{6 D_p C_A}$$

Gas-film   Pore diffusion

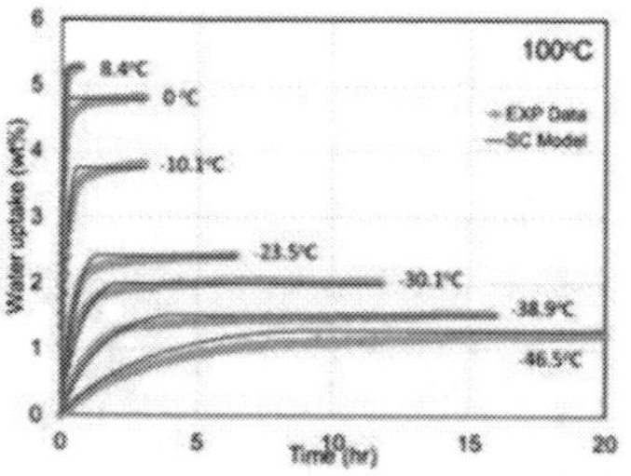

Linear Driving Force (LDF) Model:

$$\frac{d\bar{q}}{dt} = k_{LDF}\left(q_e - \bar{q}\right)$$

$$\frac{1}{k_{LDF}} = \frac{R_p}{3k_f}\frac{q_e \rho_p}{C_b} + \frac{R_p^2}{15\varepsilon_p D_p}\frac{q_e \rho_p}{C_b} + \frac{R_c^2}{15 D_c}$$

Gas-film   Macropore   Micropore

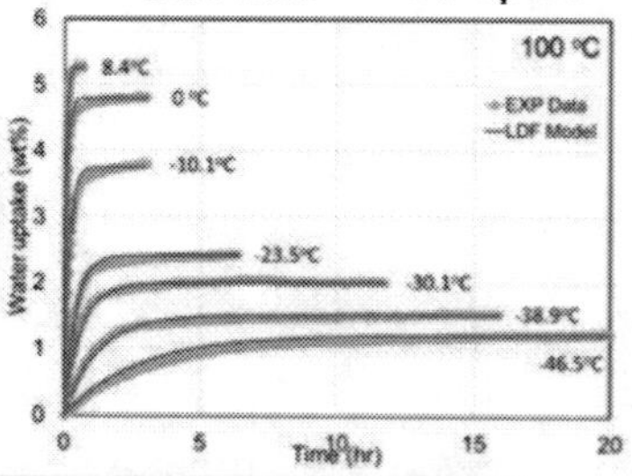

k_f= 10.4 ~ 1.76 cm/s; D_p = 0.010 ~ 0.016 cm²/s

GSTA VS H. LANGMUIR

GSTA Isotherm:

$$q = \frac{q_{\max}}{m} \frac{\sum_{n=1}^{m} n K_n p^n}{1 + \sum_{n=1}^{m} K_n p^n}$$

Parameter: q_{max}, m, $K_1 \ldots K_m$

For m=5, 7 parameters need to be determined

Heterogeneous Langmuir Isotherm:

$$q = \frac{q_{\max,1} K_1 p^1}{1 + K_1 p^1} + \frac{q_{\max,2} K_2 p^2}{1 + K_2 p^2} + \ldots\ldots + \frac{q_{\max,m} K_m p^m}{1 + K_m p^m}$$

Parameter: $q_{max,1} \ldots q_{max,m}$, $K_1 \ldots K_m$

For m=5, 10 parameters need to be determined

MODELING OF H_2O ISOTHERMS

- q_{max} = 8. 56 wt%
- m=3; Three energetically distinct adsorption sites

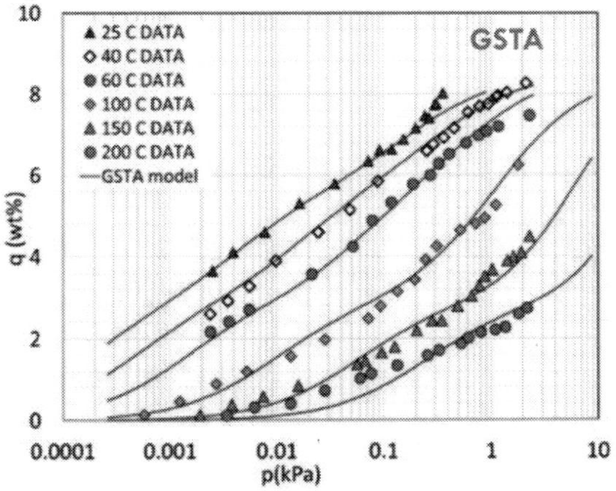

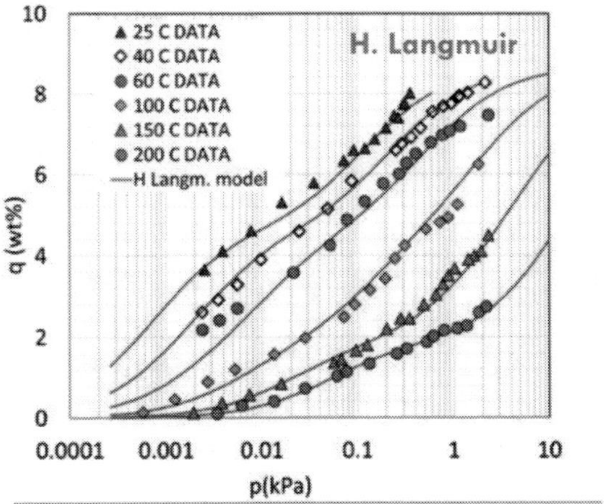

n	$q_{max,n}$ (wt.%)	ΔH° kJ/mol	ΔS° J/K/mol	R^2
1	2.85	-43.45	-74.73	0.990
2	2.85	-111.54	-250.78	0.997
3	2.85	-141.20	-328.19	0.997

n	$q_{max,n}$ (wt.%)	ΔH° kJ/mol	ΔS° J/K/mol	R^2
1	1.82	-30.52	-39.775	0.978
2	4.01	-34.86	-99.37	0.983
3	2.73	-64.24	-156.53	0.994

CO-ADSORPTION OF I_2 AND H_2O ON Ag^0Z

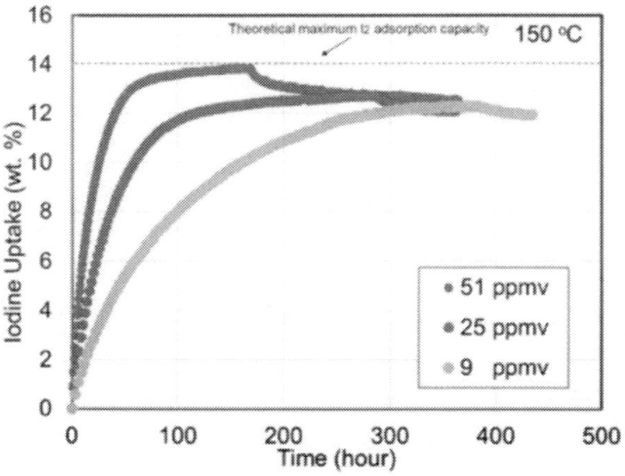

Iodine adsorption/desorption in dry air

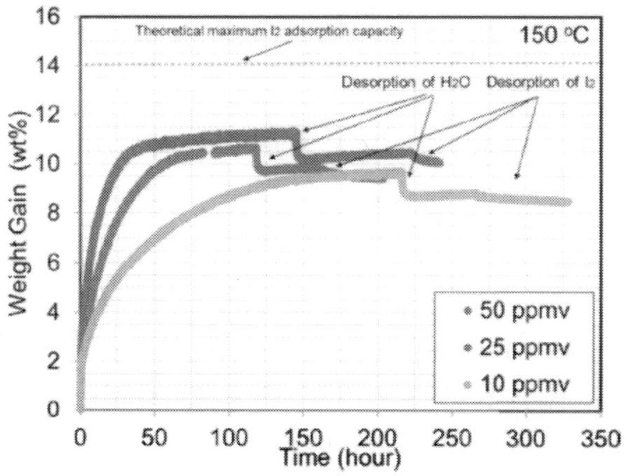

Iodine co-adsorption with H_2O/desorption in moist air (Dew point: -18 °C)

CO-ADSORPTION OF I_2 AND H_2O ON Ag^0Z

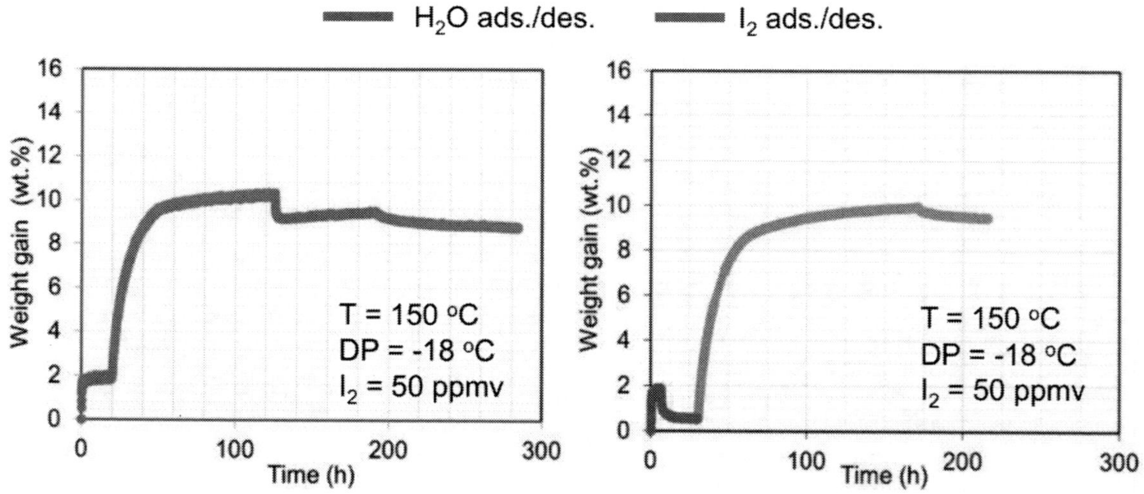

- ❑ Ag adsorption sites were deactivated by H_2O
- ❑ Mechanisms of deactivation need to be studied
- ❑ Chemical analyses in progress

CO-ADSORPTION MODELING

- Multicomponent predictions based on Generalized Predictive Adsorbed Solution Theory (GPAST) [5]
  - Provides the prediction of multi-component adsorption equilibria given the isotherm parameters and conditions
  - Optimization performed by the MGPIE Application developed at GIT.
- Modeling results with GPAST indicate:
  - GPAST cannot describe the co-adsorption of I_2 and H_2O
  - I_2 and H_2O are not sharing the adsorption sites
  - Deactivation of iodine loading sites may occur during adsorption processes

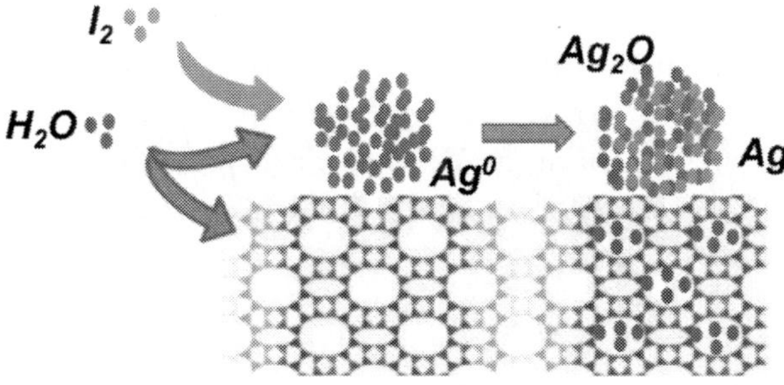

5. Ladshaw et al, *Fluid Phase Equilibria*, 2014; Ladshaw et al, *AIChE*, 2015.

EXTENDED PROJECT ON AGING ISSUES

Study the mechanisms of the aging processes of Ag containing adsorbents in off-gases.

- Ag containing adsorbents: *Ag mordenite* and *Ag functionalized aerogel*
- Off-gases: Air, H_2O vapor, NO, and NO_2
- Objective: Determine the mechanism of deactivation of Ag adsorption sites for iodine over time up to 6 months
- Supported by DOE NEUP

NEUP Nuclear Energy University Programs
U.S. Department of Energy

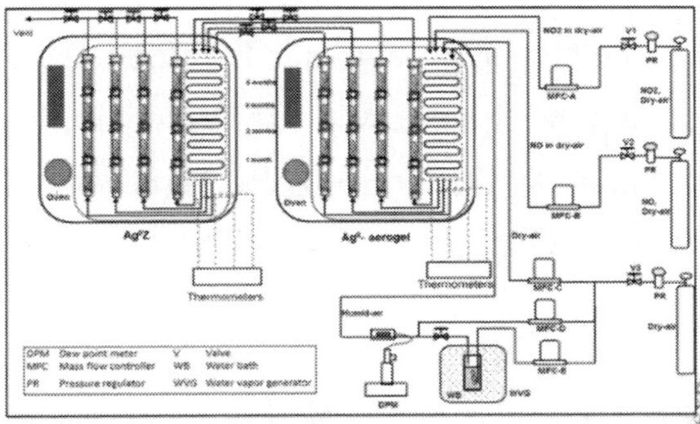

CONCLUSIONS AND FUTURE WORK

- ❑ Fundamental adsorption experiments and modeling has provided information of the adsorbent properties and parameters of adsorption kinetics and equilibria

- ❑ Co-adsorption of iodine and water can not be described by GPAST due to a co-adsorption mechanism including the deactivation of adsorption sites
  - The model will be modified, or new model will be developed for co-adsorption of iodine and water
  - Mechanism of iodine and water co-adsorption will be studied

- ❑ Aging issues of Ag containing sorbents (Ag^0Z and Ag^0-aerogel) under off-gas conditions
  - Determine the mechanisms of aging (deactivation) processes
  - Incorporate the deactivation processes to the adsorption models

ACKNOWLEDGEMENTS

Funding support

- DOE NEUP; Standard Research Contracts: NE0008275
- TPOC: Robert Jubin
- Federal Manager: John Gilligan

Collaborator

- Jack D. Law, Amy Welty, and Kevin Lyon; Idaho National Laboratory

The off-gas sorption modeling and verification team

- Yue Nan and Lawrence L Tavlarides (PI); SU
- Austin P Ladshaw, Sotira Yiacoumi (Co-PI) and Costas Tsouris (Co-PI); GIT
- David W DePaoli (Co-PI) and Costas Tsouris (Co-PI); ORNL
- Jorge Gabitto (Co-PI); PVA&M

464274 Renuclear: Use of a Molten Salt Breeder Reactor for Nuclear Reactor Decomissioning and Spent Fuel Disposal

Abstract:

Nuclear reactors provide a reliable, low-emission source of baseload energy. However, the generation of radioactive spent nuclear fuel (SNF) is a prominent obstacle to increased implementation of nuclear power. ReNUclear aims to address the inefficiencies of established nuclear technology by retrofitting existing generation II and III reactors with a new molten salt breeder reactor (MSBR) and chemical treatment system to process SNF. This endeavor, titled "Project ReNU", will provide a more economically and environmentally sound decommissioning program for nuclear reactors. Dissolving SNF in a high temperature molten salt allows for more complete fuel consumption than is possible with existing reactor technology in doing so reduces the half-life of SNF by several orders of magnitude. Our process will allow for the recovery of uranium and plutonium dioxides from the SNF which will be formed into new fuel pellets for existing generation II and III nuclear reactors. By tackling the problem of SNF storage, Project ReNU and ReNUclear will play a significant role in increasing the profitability, safety, and attractiveness of nuclear power as a source of low-carbon electricity.

458927 Analysis of Mobile Helium Cluster Dynamics Near Surfaces and Grain Boundaries of Plasma-Exposed Tungsten

Abstract:

Plasma facing components (PFCs) as the first wall materials in nuclear fusion reactors are exposed to intense plasma heat and particle fluxes. The implantation of helium (He) atoms into these materials impacts significantly their surface morphological and near-surface structural evolution. Tungsten (W) is a promising PFC material because of its thermomechanical properties. In tungsten, such interstitial He atoms are very mobile and aggregate to form clusters of various sizes. The smallest of these helium clusters, containing n He atoms with n = 1-7, also are mobile and their diffusional transport mediates the evolution of surface morphology and the sub-surface helium gas bubble structure and dynamics.

In this presentation, we report the results of a systematic atomic-scale analysis of the dynamics of small mobile helium clusters near low-Miller-index tungsten surfaces, aiming at a fundamental understanding of the near-surface transport and kinetics of helium-carrying species in plasma-exposed tungsten. These small mobile helium clusters are attracted to the surface and migrate to the surface by Fickian diffusion and drift due to an elastic interaction that generates a thermodynamic driving force for surface segregation. As the clusters migrate toward the surface, trap mutation (TM) and cluster dissociation reactions are activated at rates higher than in the bulk. TM produces W adatoms and immobile complexes of helium clusters surrounding W vacancies located within the lattice planes at a short distance from the surface. These reactions are identified and characterized in detail based on analysis of a large number of molecular-dynamics (MD) trajectories for each such mobile cluster near W(100), W(110), W(111), and W(211) surfaces. TM is found to be the dominant cluster reaction for all cluster and surface combinations, except for the clusters with n = 4 and n = 5 near W(100) where cluster partial dissociation following TM dominates. We find that there exists a critical cluster size, n = 4 near W(100), W(111), and W(211) and n = 5 near W(110), beyond which formation of multiple W adatoms and vacancies in TM reactions is observed.

In addition, we have carried out a systematic atomic-scale analysis of the dynamics of small mobile helium clusters near a prototypical, symmetric tilt grain boundary (GB) in tungsten based on MD simulations. The mobile clusters are attracted to the GB by an elastic interaction, similar to that with the surfaces, which generates a drift flux toward the GB and leads to helium segregation on the grain boundary. We find that TM reactions are activated in the vicinity of the GB, they are the dominant kinetic processes for clusters with n = 4-7, and only a single vacancy is formed in each of the TM reactions identified regardless of cluster size n. The displaced W atom generated from the TM reaction near the GB forms an extended W interstitial configuration on the GB. This interstitial configuration is characterized by mobility that depends on the location where the TM reaction occurs: it is immobile when the vacancy produced by the TM reaction is located a few lattice planes away from the GB plane and highly mobile along a specific direction when the produced vacancy is located on the GB. The latter mechanism initiates a potentially fast migration path for W atoms along the GB toward a free surface, which may influence significantly the surface morphology of plasma-exposed tungsten.

The identified cluster reactions in the vicinity of surfaces and grain boundaries in tungsten are responsible for important structural, morphological, and compositional features in plasma-exposed tungsten, including surface adatom populations, near-surface immobile helium-vacancy complexes, and retained helium content in the tungsten, which are observed in our large-space-scale MD simulations of He implantation and resulting near-surface evolution in tungsten. These features are expected to strongly influence the amount of hydrogen re-cycling and tritium retention in fusion tokamaks. We have also analyzed the effects of helium segregation on surfaces and grain boundaries in tungsten on their elastic interactions with mobile helium clusters and on the resulting cluster reaction rates. The results of our study contribute significantly to the parameterization of continuum transport-reaction models for the computationally efficient, coarse-grained modeling of such cluster dynamics.

459364 Experimental Investigation of the Pebbles Residence Time in a Pebble Bed Reactor (PBR) Using Residence Time Distribution (RTD) Technique

Abstract:

The granular flow of pebbles in a pebble bed reactor (PBR) demonstrated the flow under the influence of gravity with long lasting frictional contacts. The basic physics governing it is not fully understood and hence the dynamic core of a PBR and non-idealities associated with pebbles flow inside the reactor core are of trivial significance from safety analysis and licensing, and thermal hydraulics point of view. The experimental setup mimics the slow flow of pebbles in the core under the influence of gravity. The pebble bed test reactor made of acrylic material (1 foot outer diameter with 11.95 inch inside diameter and 1 foot in height) is filled with ½" glass marbles. The main reason to choose these dimensions of test reactor is from experimental feasibility point of view. The residence time of pebbles in the core is an important parameter from various neutronic and safety related considerations. This can be controlled by controlling the exit flow rate of pebbles and control over radial position of returned pebbles. An exit controller is installed at the bottom opening in the cone. An exit flow rate of one pebble exiting every five seconds is used which can be set to the desired exit flow rate used in PBR (one pebble exiting every thirty sec or higher). The main reason to use this exit flow rate is to avoid experiments of prolonged duration. The glass marbles coming out of opening in extractor tube, which is operated by a solenoid operated slider, falls into a conveyor bin just below the reactor. From there, the glass marble is transferred back to the hopper at the top via adjustable speed conveyor (TipTrak from UNITRAK). Conveyor bin releases glass marbles in this hopper. Marbles are then transferred to the top of the reactor via inlet control mechanism which consists of straight and elbow sections of one pebble diameter tube. This inlet control mechanism also has three swivel joints. The inlet control mechanism is connected to a top plate (diameter matching with the reactor) having 17 holes. These holes are provided to return the pebble at 17 different radial positions in a non-violent manner. The vertical leg of conveyor belt is kept at a sufficient distance (~150 cm) away from the test

reactor.

In the current study, investigations of overall pebbles residence time are carried out by implementing radioisotopes non-invasive flow visualization techniques such as residence time distribution (RTD) measurement technique. A dedicated RTD setup consisting of two collimated scintillation detectors was implemented around the continuous pebble re-circulation experimental setup. One detector located at the entrance of test reactor and the other one at the exit reactor. A lead collimator for scintillation detectors is fabricated using water-jet machining facility available with Missouri S&T. The slit in the collimator is 2" in length, 1" thick and has a width of 1 mm. The tracer position is identified by simultaneously monitoring photo-peak counts received by these detectors. In the current setup design, the detector collimators are having horizontal slits. When the radioactive tracer particle is in the plane of horizontal slit, maximum counts are recorded. This principle is used to record the time of entry and departure of tracer particle from which overall residence time of pebbles is calculated for different initial seeding positions of tracer particle. In addition, this techniques can provide further insights on non-idealities, i.e. stagnant zones, associated with pebbles flow in the core in a non-invasive manner. A stagnant/dead zone may exist in the pebble bed reactor near the transition from cylindrical to conical section at the pebbles outlet section for recirculation from the bottom of PBR as shown experimentally in Gatt's study. The results show the characteristics of overall pebble residence time/transient number at different initial seeding positions of a tracer particle. It has been reported that the overall pebbles residence time/transient increases with change in dimensionless initial seeding position (r/R) from the center towards the wall (169 % increase is observed for r/R of 0.92 with respect to r/R of 0). Residence time experimental results provided benchmark data that could be used for assessment of commercial Computational Fluid Dynamics (CFD) codes such as Discrete Element Method (DEM) based simulation results. These validated computational methodologies can then be used to carry out high fidelity simulations and obtain predictions of actual scale PBR core dynamics.

470494 Screening of Polymeric Materials for Tritiated Water Separation By Molecular Simulations

Abstract:

Separation of tritiated water (HTO) from nuclear power plants is extremely challenging because the thermophysical properties of HTO and H_2O *are so similar. Hence, there is a critical need for developing new separations technologies that are energetically and economically efficient. It has been experimentally observed that some polymers can be effective at separating water isotopologues but the mechanism is not completely understood. Our hypothesis is that the difference in the strength of hydrogen bonds between the water isotoplogues and polar groups on the polymer is responsible for the observed separation. We use molecular simulations employing path integral molecular dynamics (PIMD) to investigate the effectiveness of different candidate polymers for isotope separation. The lifetime of hydrogen bonds between water isotopologues and the polymers is used as a metric for the efficiency of a given candidate polymer towards better materials. Effects of temperature and concentration of tritiated water towards the separation will be discussed. Machine-learning is used to study the hydrogen-bonding network in bulk water and water within the polymer.*

457916 Comparison Between the OLI-MSE and eNRTL Models in Predicting Thermodynamic Properties of the $NaNO_3$-HNO_3-H_2O Ternary System

Abstract:

Chemical engineers engaged in spent nuclear fuel and nuclear waste processing have pursued high fidelity process modeling and simulation as have their counterparts in the conventional chemical industries. Although current modeling and simulation is capable of representing the physical and chemical processes in nuclear waste treatment at a certain level of sophistication, accurate thermodynamic models for the systems involved are essential not only to predict their chemical and physical properties correctly, but also to support high fidelity process simulation. Since reprocessing usually involves extraction with concentrated nitric acid followed by neutralization with caustic soda (e.g., the PUREX process), large amounts of sodium nitrate are formed. Consequently, three compounds – nitric acid, sodium nitrate, and water – are typically the most abundant species in nuclear waste liquids. The OLI-MSE (mixed solvent electrolyte) and eNRTL (electrolyte non-random two-liquid) are two of the activity coefficient models most widely used in the nuclear industry for aqueous nitric acid solutions. In the present work, the two models are compared side-by-side for the first time by having their results checked against literature experimental data on vapor-liquid equilibrium, calorimetric properties (i.e., heat capacity and liquid molar enthalpy), and speciation for the $NaNO_3$-HNO_3-H_2O ternary system and its subsystems.

462300 Multi-Agent Optimization Framework (MAOP) for Synthesizing Optimal Radioactive Waste Blends

Abstract:

Multi-agent Optimization Framework (MAOP) for synthesizing optimal radioactive waste blends

Berhane H. Gebreslassie and Urmila M. Diwekar

Center for Uncertain Systems, Tools for Optimization and Management (CUSTOM):Vishwamitra Research Institute, Crystal Lake, IL 60012 â€" USA

berhane@vri-custom.org; urmila@vri-custom.org

Abstract

The multi-agent optimization (MAOP) framework provides a way of combining various algorithms in one platform and exploits the strength possessed by each algorithm. Most of large scale process system engineering problems include nonlinear and non-convex problems. In this work, we propose a multi-agent optimization framework for solving complex large scale process system engineering problems. The framework uses a variety of different algorithmic agents which include the gradient based local optimizers and metaheuristic algorithms (efficient simulated annealing, efficient genetic algorithm and efficient ant colony optimization algorithms). Each agent encapsulates a particular problem-solving procedure. We investigate the effect of cooperation among agents of the multi-agent system combined into a framework designed to solve large scale combinatorial optimization problems. Computational experiments are carried out using benchmark problems and real world case study. The proposed methodology enables to improve the quality of solutions and the computational efficiency in comparison with non-cooperative multi-agent framework and standalone agents. The MAOP framework which includes the five major parts of the algorithm (figure 1) is given in the figure 2.

Fig. 1. The major parts of the MAOP algorithm and the information flow direction.

Fig. 2. The basic flow diagram of the heterogeneous MAOP algorithm

Case Study:

A real-world problem of synthesizing optimal waste blends is analyzed to test the applicability of this novel approach in addressing a general synthesis problem. The Hanford Site in southeastern Washington produced nuclear materials using various processes for nearly fifty years. Radioactive waste was produced as by-products of the processes. This waste can be retrieved and separated into high-level and low-level wastes and subsequently immobilized for future disposal. The high-level waste is converted into a glass form before disposal. However, the glass must meet both process-ability and durability restrictions. The process-ability conditions ensure that during processing, the glass melt has properties such as viscosity, electrical conductivity and liquidâ€™s temperature such that they lie within the range of known to be acceptable for the vitrification process. Durability restrictions ensure that the resultant glass meets the quantitative criteria for disposal in a repository. There are also bounds on the composition of the various components in the glass. The radioactive waste and frit are mixed and heated to form a glass that satisfies the constraints related to the glass property and configuration.

Hanford has 177 tanks (50,000 to 1 million gallons) containing radioactive waste. These wastes vary widely in composition, and the glasses produced from these wastes will be limited by a variety of components. The minimum amount of frit would be used if all the high-level wastes were combined to form a single feed to the vitrification process. Because of the volume of waste involved and the time span over which it will be processed, this is logistically impossible. However, much of the similar benefit can be obtained by forming blends from sets of tanks. The problem is how to divide all the tanks into sets to be blended together such that the minimal amount of frit is used. Therefore, the goal of this work is to find the optimal blend configuration that minimizes the amount of frit added to convert the radioactive waste into glass. Minimizing the amount of frit added to a minimum amount has two explicit reasons.

Fig. 3. Processing of radioactive wastes to glass

References

1. Siirola JD, Steinar Hauan S, and Westerberg AW, 2003. Toward agent-based process systems engineering: proposed framework and application to non-convex optimization. *Computers and Chemical Engineering* 27: 1801-1811.
2. Gebreslassie BH, and Diwekar UM, 2015. Efficient ant colony optimization for computer aided molecular design: Case study solvent selection problem. *Computers and Chemical Engineering* 78: 1-9.
3. Diwekar UM, Xu W, 2005. Improved genetic algorithms for deterministic optimization and optimization under uncertainty. part I. algorithms development. *Industrial and Engineering Chemistry Research* 44(18) 7132-7137.
4. Venkatesh Narayan V, Diwekar UM, Hoza M. Synthesizing Optimal Waste Blends. *Computers chem. Engng* Vol. 20, Suppl., pp. SI443-SI448, 1996.
5. Chaudhuri P, Diwekar UM, 1999. Synthesis Approach to the Determination of Optimal Waste Blends under Uncertainty. *AIChE Journal*. 45: 1671-1687.

462734 Thermodynamics Analysis and Optimization of a Stirling Cycle for Lunar Surface Nuclear Power System

Abstract:

A nuclear powered stirling engine system in lunar surface composed of a reactor core, a hot heat exchanger, a stirling engine, a cold heat exchanger and a radiator. A mathematical model for the overall thermal efficiency of the lunar surface nuclear powered stirling engine with the heat loss of the hot heat exchanger, finite rate heat transfer, regenerative heat loss, finite regeneration process time and conductive thermal bridging losses is developed. The optimal efficiency of the system was determined by the finite time thermodynamics. The effect of some system parameters such as the absorbed temperature, the heat rejection temperature, the convection heat transfer coefficients of the hot and cold heat exchangers on the optimal efficiency were analyzed in details. The results showed that the overall thermal efficiency increased and then decreased with the absorbed temperature increase. The highest thermal efficiency was about 28%, which was far away from the corresponding carnot efficiency. Higher rejection temperature had bad effect on the thermal efficiency but could improve the waste heat rejection to the outer space. Enhancement of the heat transfer could increase the thermal efficiency. The results obtained here may provide a new ideal to design lunar surface nuclear powered stirling cycle.

468396 Simulation of Helium Transport Near Prismatic Dislocation Loops in Tungsten

Abstract:

We analyze the effect of sub-surface prismatic dislocation loops on the surface morphology and helium clustering behavior of plasma-facing tungsten through the use of molecular dynamics simulations. These simulations are moderately large, 25 nm x 25 nm x 24 nm, consisting of approximately 830,000 atoms and simulated to times on the order of one microsecond. This approach eliminates some finite-size effects common in smaller simulations and reduces the flux to approximately 5.5 x 10^26/m^2/s, including ions that reflect back into the plasma. This flux is approximately a factor of 15 lower than is typically used in smaller simulations. These results indicate that prismatic loops with radii of approximately 3 nm that are centered 10 nm below the surface with Burgers vectors parallel to the surface initially cause helium atom clusters to accumulate at the edges of the dislocation core relatively quickly (within 100-150 ns of the onset of plasma exposure). Subsequent growth of these clusters, however, is relatively minimal. This is partially explained by the relatively high helium implantation flux, which causes bubbles to accumulate 0-7 nm below the surface and "block" the region of the metal containing the dislocation. Another effect results from the strain field around the loop itself: the compressive regions along the direction of the Burgers vector repel helium, but the tensile region initially attracts helium and traps it. However, we think that the attractive tensile stress region is effectively shielded by the formation of helium clusters above it, and these bubbles subsequently experience relatively slow growth.

468183 Atomistic and Continuum Drift-Diffusion Simulations of Helium Transport in Tungsten

Abstract:

Helium from linear and tokamak-like plasma devices has been known to cause "fuzz" and other nanometer-scale features on the surface of tungsten and other metal surfaces after a few hours of plasma exposure. The mechanisms involved occur on multiple length and time scales, and understanding all of those details is the subject of significant multi-scale modeling efforts. In particular, the spatial and temporal scales necessary to simulate even a small portion of the tungsten divertor in a real fusion device make atomistic simulations intractable, so we have sought a continuum-based description that is informed by atomistic simulations of more accessible systems. This effort comes in four parts: (1) large-scale atomistic simulations of helium diffusion in tungsten, which aims to discover mechanisms and provide a benchmark against which more coarse-grained models can be tested; (2) small, targeted molecular dynamics and statics calculations that quantify the rates of specific mechanisms, such as drift forces near surfaces and grain boundaries, and (3) coarse-grained, drift-reaction-diffusion models that incorporate information from small-scale molecular dynamics simulations and use large-scale simulations as a benchmark for still-larger, more realistically-sized simulations of plasma-facing tungsten. The orientation of the surfaces, the presence of grain boundaries, and the presence of other bubbles all have a strong impact on the rate of local helium transport, resulting in a spatially-dependent rate of helium transport in the divertor. These tools aim to provide a tractable and reasonably accurate discription of helium bubble distributions over time, an important step in the process of understanding tritium retention and material fatigue in nuclear fusion reactors.

472779 Density Functional Theory Study of the Hydrolysis of a Sodium Borosilicate Glass

Abstract:

The long-term stability of nuclear waste glass forms is difficult to predict, and estimates rely on empirical information from natural analogs or accelerated tests. Models informed by first-principles calculations have the potential to improve the description of long-term behavior based on atomic-scale details. In this contribution, we present reaction barriers and energies for the hydrolysis of a model sodium borosilicate glass calculated from density functional theory considering acid, base, and neutral conditions. We discuss the roles of the glass composition and structure on glass stability in potential environmental conditions.

PORE SATURATION MODEL FOR CAPILLARY IMBIBITION AND DRAINAGE PRESSURES

Dr. James E. Laurinat
Savannah River National Laboratory
Aiken, South Carolina

Abstract

Leaching and transport of radionuclides from cementitious waste forms and from waste tanks is a concern at the Savannah River Site and other Department of Energy sites. Computer models are used to predict the rate and direction for migration of these through the surrounding soil. These models commonly utilize relative permeability and capillary pressure correlations to calculate migration rates in the vadose (unsaturated) zone between the surface and the water table. The most commonly used capillary pressure models utilize two parameters to relate the pressure to the relative saturation between the wetting (liquid) and nonwetting (gas) phases. The correlation typically takes the form of a power law relation (e.g., Brooks and Corey and van Genuchten)[1,2] or an exponential equation (e.g., Kosugi).[3,4]

In this study, a pore saturation model is used to derive the capillary pressure as a function of a characteristic pore pressure and the wetting phase saturation. Singularity analyses of the total energies of the wetting and nonwetting phases give residual saturations for the two phases. The model includes separate pressures for imbibition and drainage to account for capillary hysteresis.

The model successfully correlates a selected set of laboratory imbibition and drainage data for sand, from the U. S. Salinity Laboratory's UNSODA database[5,6] and other sources.

The capillary pressure model utilizes a single fitting parameter, a characteristic pore pressure, which is related to a characteristic pore diameter by the Laplace equation. Data regression shows that this pore diameter approximately equals the diameter predicted from a capillary rise analysis based on the mean particle diameter.

Nomenclature

$\bar{d}_{part}$	volume average particle diameter for porous material
d_{pore}	effective pore diameter for porous material
G_s	potential energy associated with saturation of the porous medium
ΔG_{jump}	Gibbs free energy associated with jump condition between a homogeneous porous material and a porous material with fissures
H	molar enthalpy
k	Darcy coefficient for permeability
k_n	relative permeability to the nonwetting phase
k_w	relative permeability to the wetting phase
n_i	number of moles of the ith solute in the Gibbs' adsorption theorem calculation
P_a	average pressure in the pores of the porous medium
P_c	intrinsic capillary pore pressure, equal to the gas-liquid pressure difference

$P_{c,d}$ capillary pressure measured for drainage of the wetting phase

$P_{c,d,ms}$ capillary pressure measured for metastable drainage of the wetting phase

$P_{c,i}$ capillary pressure measured for imbibition of the wetting phase

$P_{c,i,0}$ minimum capillary pressure required to imbibe the wetting phase at the residual nonwetting phase saturation

$\Delta P_{c,ov}$ maximum reversible overshoot in the capillary pressure at the minimum saturation

P_{flow} characteristic pressure for flow in capillary pores

P_g gas phase pressure in capillary pores

ΔP_{jump} pressure change associated with jump condition between a homogeneous porous material and a porous material with fissures, applied to the liquid phase

P_l liquid phase pressure in capillary pores

P_{pot} potential pressure associated with partial saturation of capillary pores

$P_{v,i}$ vapor pressure of the ith solute in the Gibbs' adsorption theorem calculation

P_o ambient pressure for liquid in contact with the porous medium

R_g universal gas law constant

s relative saturation with the wetting phase

s^* saturation for a fissured soil, scaled so that it gives a capillary pressure equal to that of a nonfissured soil

s_{cr} critical saturation where the capillary pressure for imbibition equals the capillary pressure for drainage

s_{max} maximum saturation for a fissured porous material

s_{min} minimum saturation for a fissured porous material

s_{nr} residual nonwetting phase saturation

s_{wr} residual wetting phase saturation

T absolute temperature

v_n velocity of the nonwetting phase

v_w velocity of the wetting phase

W work required for flow through the porous medium

W_g work required for flow of gas through the porous medium

W_l work required for flow of liquid through the porous medium

z distance in the direction of flow

ε porosity

θ wetting angle

μ_n dynamic viscosity of the gas or nonwetting fluid

μ_w dynamic viscosity of the liquid or wetting fluid

σ air-water interfacial tension

Introduction

Capillary pressure represents the pressure required to displace a given volume of one of the phases from a two-phase mixture in a porous medium. The capillary pressure is typically measured either by imbibition or by drainage. Imbibition pressure measurements begin with uptake of liquid by the porous medium, typically a sand or soil, which initially is saturated with gas. For drainage pressure measurements, the gas displaces the liquid. (For convenience, this paper will refer to the terms wetting phase and liquid and nonwetting phase and gas interchangeably.)

During drainage, there exists a residual wetting saturation, s_{wr}, at which no additional amount of liquid can be displaced, no matter how much pressure is applied. For imbibition, as the liquid saturation increases, the imbibition pressure decreases. The pressure drops to zero, and absorption of liquid stops, at a maximum fractional wetting phase saturation, s_{nr}, called the residual nonwetting saturation. Significant drainage does not occur until a threshold pressure referred to as the displacement pressure or the entry head is exceeded. The drainage pressure then increases as the liquid saturation decreases. At any given saturation, the drainage capillary pressure exceeds the imbibition capillary pressure. The hysteresis between the two pressures is greatest at the residual nonwetting saturation.

Historically, the capillary pressure has been correlated as a function of the liquid saturation, normalized with respect to the difference between the residual nonwetting and wetting saturations. Most capillary pressure models utilize two parameters to relate the pressure to the relative saturation. The relationship typically takes the form of a power law relation (e.g., Brooks and Corey and van Genuchten),[1,2] an exponential equation (e.g., Kosugi),[3,4] or a combination of these two forms applied over different liquid saturation ranges (e.g., Rossi and Nimmo).[7] Brooks and Corey assigned the entry head, defined as the minimum capillary pressure during drainage, as one parameter and used the other parameter to characterize the pore size distribution. Several other models have followed this convention.

Parker and Lenhard[8] addressed the hysteresis between drainage and imbibition. They developed a model that includes effects of gas entrapment during imbibition. Their model defines limiting maximum capillary heads for drainage and minimum capillary heads for imbibition. They state that the actual pressure must lie at some value between these limits that depends on the flow history. A simple method of applying this model, applied by White and Oostrom,[9] is to define separate limiting capillary pressures for imbibition and drainage.

This paper presents a departure from traditional analytic models. A probabilistic pore pressure model and global energy balances replace the semi-empirical approach of other models. The energy balances are solved for singularities in the driving forces for flow of the wetting and nonwetting phases, to obtain the residual wetting and nonwetting saturations, respectively. The pore pressure models are then used to calculate the capillary pressure as a function of saturation. Results are compared to selected capillary pressure measurements.

Derivation of Energy Balance for Capillary Pressure Model

The singularity analyses for the residual saturations are based on comparisons of the potential energy associated with the difference between the pressures of the two phases in the porous medium and the work required for flow of each phase. The analysis is loosely analogous to the Gibbs' adsorption theorem, which relates work done by surface forces to the chemical potential of a system. The model replaces the chemical potential with a mechanical potential based on the stored interfacial energy and substitutes a flow work term for the work needed to extend an interface. The model is applied to an adiabatic control volume within a larger volume of the porous medium, from which the only energy transfer occurs by flow of gas and liquid in or out. The model evaluates changes of the potential energy within the control volume and changes in the amount of flow work performed by the control volume on its surroundings.

The energy balance for the control volume takes the form

$$\Delta H = \Delta G_S + \Delta W \quad (1)$$

The energy balance is formulated in terms of the enthalpy because it is an open system in which liquid and gas may cross the volume boundaries.

Since the model assumes that the porous medium control volume is adiabatic, the enthalpy term is zero, and

$$\Delta G_S = -\Delta W \quad (2)$$

The mechanical potential and flow work terms can be expressed in terms of pressures much as the chemical potential in the Gibbs' adsorption equation. For a control volume initially comprised of one mole of gas, these expressions are

$$\Delta G_S = -R_g T \Delta \ln\left(P_{pot}\right) \quad (3)$$

and

$$\Delta W = -R_g T \Delta\left(-\ln\left(P_{flow}\right)\right) \quad (4)$$

The model employs different flow pressures for imbibition and drainage. The flow pressure for imbibition is based on liquid phase flow, while the flow pressure for drainage is based on gas phase flow.

Derivation of Pore Pressure Model

To derive the pressures for the potential energy and the flow work, a model for the pore pressure as a function of saturation must be developed. The pore pressure model assumes that each pore is filled with either a liquid wetting or a gaseous nonwetting phase. The capillary pore pressure is distributed by gas-liquid interfaces between the pores. The distribution of gas-and liquid-filled pores and the connections between pores are assumed to be random. The difference between the average pore pressure and the liquid pressure, according to this model, is equal to the product of the capillary pore pressure and the probability that a liquid-filled pore is in contact with a gas-filled pore and not with liquid in another pore. This probability, in turn, is the gas saturation. Thus, the difference between the average pore pressure and the liquid pressure is given by

$$P_a - P_l = P_c\left(1 - s\right) \quad (5)$$

Likewise, the difference between the gas pressure and the average pore pressure is equal to the product of the capillary pressure and the liquid saturation, or,

$$P_g - P_a = P_c s \tag{6}$$

The potential energy is calculated from the difference between the average pressure in the porous medium, P_a, and the reference pressure for flow through the porous medium, P_o. For an influx of liquid into the porous medium (imbibition), the reference pressure is the liquid phase pressure, and

$$P_a - P_o = P_c(1-s) \tag{7}$$

For flow of gas (drainage), the reference pressure is the gas phase pressure, so that

$$P_a - P_o = -P_c s \tag{8}$$

These pressure differences apply only to the fractions of the pore volume occupied by liquid and gas, respectively. Consequently, for either liquid or gas flow, the potential energy is based on a pressure differential given by

$$P_{pot} = P_c s(1-s) \tag{9}$$

The pressure differential is defined to be positive for both imbibition and drainage. Substitution of this pore pressure in the equation for the potential energy yields

$$\Delta G_s = -R_g T \Delta ln\left(P_c s(1-s)\right) \tag{10}$$

Analysis of Residual Saturations

At the residual wetting phase saturation, s_{wr}, the resistance to flow of the liquid becomes infinite, and, in an overall energy balance, the resistance to flow of the gas can be ignored. It may be stated, then, that for a given change in the saturation the change in the potential energy counteracts the change in the work needed to cause the liquid to flow. In other words, at $s = s_{wr}$,

$$\frac{dG_s}{ds} = -\frac{dW_l}{ds} \tag{11}$$

The liquid work function is derived from the Darcy equation for flow of two immiscible phases. The Darcy equation for the wetting phase is:

$$\varepsilon s v_w = -\frac{kk_w}{\mu_w}\frac{dP_l}{dz} = -\frac{kk_w}{\mu_w}P_c\frac{ds}{dz} \tag{12}$$

or

$$v_w = -\frac{kk_w}{\mu_w \varepsilon} P_c \frac{d\, ln(s)}{dz} \tag{13}$$

From this expression the pressure acting on a given volume of liquid is seen to be $P_c\, ln(s)$. The change in the corresponding work function is given by

$$\Delta W_l = -R_g T \Delta ln(-P_c\, ln(s)) \tag{14}$$

As with the potential energy, this expression gives the change in the work function for small changes in the saturation.

Substitution of this work term and the change in the potential energy in the energy balance yields

$$-R_g T \frac{d\, ln(P_c s(1-s))}{ds} = -R_g T \frac{d\, ln(-P_c\, ln(s))}{ds} \tag{15}$$

or

$$ln(s) = \frac{s-1}{1-2s} \tag{16}$$

This equation is satisfied for $s_{wr} = 0.236$.

A similar line of reasoning is used to calculate the residual saturation of the gas, s_{nr}. Here, the resistance to the gas flow is infinite, and the resistance to flow of the liquid can be ignored. Thus,

$$\frac{dG_s}{ds} = -\frac{dW_g}{ds} \tag{17}$$

The Darcy equation for the gas phase is:

$$\varepsilon(1-s)v_n = -\frac{kk_n}{\mu_n}\frac{dP_g}{dz} = -\frac{kk_n}{\mu_n} P_c \frac{ds}{dz} \tag{18}$$

or

$$v_n = -\frac{kk_n}{\mu_n \varepsilon}\left(-P_c \frac{d\, ln(1-s)}{dz}\right) \tag{19}$$

From this expression the pressure acting on a given volume of gas is seen to be $-P_c\, ln(1-s)$. The capillary pressure is subtracted from this pressure. The capillary pressure is supplied to the gas to convert it from a stationary state in which it is in contact with a pore wall to a mobile state in which it is surrounded by liquid. The total pressure supplied to the gas phase is, therefore, $-P_c\left(1+ln(1-s)\right)$, and the corresponding change in the work function is:

$$\Delta W_g = -R_g T \Delta ln\left(-P_c\left(1+ln(1-s)\right)\right) \tag{20}$$

Substitution of this work term and the change in the potential energy in the energy balance gives

$$-R_g T \frac{d\, ln\left(P_c s(1-s)\right)}{ds} = -R_g T \frac{d\, ln\left(-P_c\left(1+ln(1-s)\right)\right)}{ds} \tag{21}$$

or

$$ln(1-s) = \frac{3s-1}{1-2s} \tag{22}$$

This equality is satisfied for $s_{nr} = 0.884$.

Measured residual saturations for a uniform sand[10,11,12] agree almost exactly with the derived values; the measured wetting saturation, estimated by graphical interpolation, was 0.230, and the measure wetting saturation was 0.884.

Calculation of Capillary Pressures for Imbibition and Drainage

The following section describes a capillary pressure model for imbibition and drainage, based on the residual nonwetting saturation. For imbibition, the wetting phase, or liquid, supplies the motive force for displacement of the nonwetting, or gas, phase. Consequently, it may be argued that the effective capillary pressure within the porous medium is the integral of the pressure gradient for the liquid phase. During imbibition, the force to displace the gas is applied over the entire surface of the porous medium, but the medium can be considered to be aerated so that the air within the capillary pores remains at atmospheric pressure. A hydraulic advantage accompanies the force applied to the liquid in the pores, then, in inverse proportion to the liquid saturation. It follows that the measured liquid pressure is the integral of the liquid phase pressure gradient, divided by the saturation. This gives for the measured capillary pressure for imbibition:

$$\frac{P_{c,i}}{P_c} = \frac{P_{c,i,0}}{P_c} - \frac{ln(s)}{s} \tag{23}$$

where $P_{c,i,0}$ represents the minimum pressure required for imbibition of liquid into a saturated porous medium containing non-displaceable gas.

During drainage, both liquid and gas displace the liquid that leaves the porous medium. Hence it may be argued that the effective capillary pressure is the volume average of the integrals of the pressure gradients for the gas and the liquid. Again, because the air remained at atmospheric pressure during drainage, there is a hydraulic advantage applied to the measured pressure, so the effective pressure is this volume-average pressure integral, divided by the saturation.

At the maximum residual nonwetting phase saturation, the measured suction pressure drops by a step change from its value for drainage to its value for imbibition. This pressure change can be attributed to the change from a continuous gas phase at atmospheric pressure to a continuous liquid phase for which, at static equilibrium, the outside pressure equals the liquid phase pressure. Since the liquid phase pressure is less than the gas phase pressure by the characteristic capillary pore pressure, the suction pressure for imbibition must be less than the pressure for drainage by just this pressure, divided by the saturation to account for the effective hydraulic advantage. These considerations, with adjustments to account for differences between the integration constants for the liquid and volume-average pressure gradients, give, for the capillary pressure for drainage,

$$\frac{P_{c,d}}{P_c}=\frac{P_{c,i,0}}{P_c}+\frac{1}{s_{nr}}-\frac{ln(s_{nr})}{s_{nr}}+\frac{(1-s)(1+ln(1-s))-s\,ln(s)}{s}-\frac{(1-s_{nr})(1+ln(1-s_{nr}))-s_{nr}\,ln(s_{nr})}{s_{nr}} \tag{24}$$

The minimum suction pressure required for imbibition into a liquid-saturated porous medium, i.e., the entry head, most likely is a function of the capillary pore pressure. The most plausible explanation that fits the measured data is that an excess capillary force is required to displace air from an array of pores located on the surface in all three directions (one perpendicular to the surface and two in transverse directions). The required force is the three-dimensional vector sum of the forces required for unidirectional displacement, one component of which corresponds to the pressure. According to this interpretation, the minimum capillary pressure is the pore pressure multiplied by the square root of three:

$$\frac{P_{c,i,0}}{P_c}=\sqrt{3} \tag{25}$$

This entry head is applied at the outer surface of the porous medium, so this pressure difference is not normalized with respect to the liquid saturation.

A combination of the capillary pressure equations for imbibition and drainage yields a critical saturation where these two pressures coincide:

$$\frac{1-(1-s_{nr})(1+ln(s_{nr})+ln(1-s_{nr}))}{s_{nr}}=-\frac{(1-s_{cr})(1+ln(s_{cr})+ln(1-s_{cr}))}{s_{cr}} \tag{26}$$

Substitution of s_{nr} in this expression gives $s_{cr}=0.301$.

Below this saturation, the suction pressure that develops during drainage is less than the suction pressure that accompanies imbibition. The logical conclusion is that the porous medium will not drain to a saturation below this critical value, unless the source of liquid is removed from contact with the porous medium or there is an external gas flow.

Comparison of Model Predictions to Capillary Pressure Data

Data from the UNSODA database maintained by the George E. Brown, Jr., Salinity Laboratory of the U. S. Department of Agriculture,[5,6] was selected for comparison with the model. The comparison is limited to data sets with paired imbibition and drainage pressures that exhibited a significant entry head $P_{c,i,0}$ and that included particle size measurements. There were three such data sources, namely Shen and Jaynes,[13] Stauffer and Dracos,[14] and Poulosavassilis.[15] An additional data source not included in the UNSODA data base, that of Smiles et al. and Vachaud and Thony,[16,17] also is analyzed. All of these data sources contained measurements made in sand. The fitting procedure entailed a calculation of the capillary pore pressure P_c, computed as the average of the measured capillary pressures divided by the capillary pressures predicted by the model at the relative saturations for each measurement. To eliminate measurements near saturation with either liquid or gas, the calculation was limited to relative liquid saturations between 0.4 and 0.75.

Figure 1 compares these data with model predictions. As this figure shows, the Shen and Jaynes data and the Stauffer and Dracos data have minimum measured saturations about half the predicted critical minimum saturation s_{cr}, while the Poulovassilis and Smiles et al. data have minimum measured saturations approximately equal to s_{cr}.

The model fits the data closely for the Poulovassilis and Smiles et al. data. The most logical explanation for the apparent discrepancy between the minimum measured saturations for the Shen and Jaynes and the Stauffer and Dracos data and the model prediction is that the porous material develops fissures that fill with gas during drainage. According to this interpretation, the fissures are larger than the pores by a sufficient magnitude that they do not offer any flow resistance but rather simply conduct fluid among the pores. The minimum saturation for a fissured material can be derived by applying a jump condition for the development of fissures. Following the logic used in the analysis of residual wetting saturations, the pressure applied to the liquid phase across such a jump, ΔP_{jump}, is

$$\Delta P_{jump} = P_c\left(\left(1-s_{min}\right)-\left(1-s_{cr}\right)\right) = P_c\left(s_{cr}-s_{min}\right) \qquad (27)$$

The minimum saturation is the saturation that will minimize the free energy associated with this jump, ΔG_{jump}. This free energy, evaluated after the appearance of the fissures, is

$$\Delta G_{jump} = -R_g T \ln\left(P_c s_{min}\left(s_{cr}-s_{min}\right)\right) \qquad (28)$$

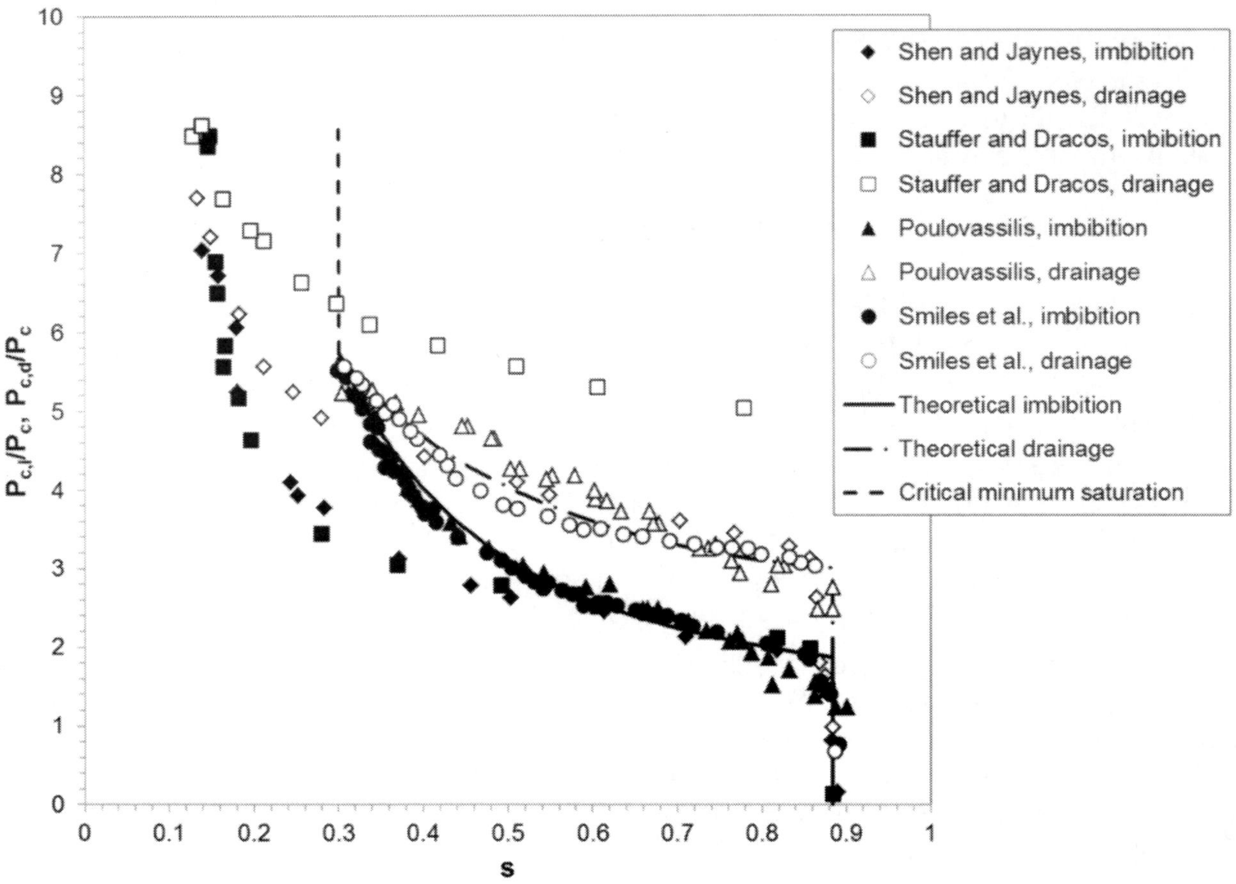

Figure 1. Variation of Imbibition and Drainage Pressures with Relative Saturation

This condition is satisfied by

$$s_{min} = 0.5s_{cr} \tag{29}$$

A similar analysis, combined with the observation that the maximum saturation for a homogeneous porous material is the residual nonwetting saturation, gives for the maximum saturation for a fissured porous medium s_{max},

$$s_{max} = 0.5 + 0.5s_{rn} \tag{30}$$

The capillary pressure model describes the saturation for just the pores of a fissured material. Because the fissure volume merely conducts pressure among the pores, the overall saturation for the total volume of porous material is a linear function of the pore saturation. If the maximum saturation is defined to be equal to the residual nonwetting saturation to conform with the model, then a linear interpolation between this saturation and the minimum saturation yields the following expression for the overall saturation of a fissured material, s^*, as a function of the model saturation for a homogeneous porous material, s, the residual nonwetting saturation, s_{nr}, and the critical saturation, s_{cr}:

$$s^* = \frac{0.5 s_{cr} s_{nr} - s_{cr} s + s_{nr} s}{s_{nr} - 0.5 s_{cr}} \tag{31}$$

Figure 2 compares the capillary pressure model with the Shen and Jaynes data and the Stauffer and Dracos data defined by s^*. The model fits the adjusted imbibition data closely, but the Stauffer and Dracos drainage data deviate significantly from the model. The difference between the measured and predicted drainage pressures remains approximately constant as the saturation increases. This suggests the presence of a metastable equilibrium that originates from the upstream (dry) side of the capillary pressure gradient.

The most plausible explanation for the overshoot and apparent metastability phenomena is that at the minimum liquid saturation, the interstitial gas in the fissures begins to exert pressure on the residual liquid in the pores. As explained in the capillary pressure model derivation, this pressure would be the product of the capillary pore pressure P_c and the fraction of the volume occupied by the interstitial gas, $0.5(1-0.5s_{cr})$, divided by the residual pore saturation, $0.5s_{cr}$. Thus, the overshoot pressure $\Delta P_{c,ov}$ is given by

$$\frac{\Delta P_{c,ov}}{P_c} = \frac{1-0.5s_{cr}}{s_{cr}} \tag{32}$$

The apparent metastability, according to this interpretation, results when the overshoot in the capillary pressure at minimum saturation propagates downstream in the nonwetting (gas) phase. The fraction of the overshoot pressure that propagates would be equal to the fraction of the pore volume occupied by gas, $1-s_{cr}$. If this overshoot pressure propagates, then the metastable drainage pressure $P_{c,d,ms}$ would be given by

$$\frac{P_{c,d,ms}}{P_c} = \frac{P_{c,d}}{P_c} + \frac{(1-s_{cr})(1-0.5s_{cr})}{s_{cr}} \tag{33}$$

As shown by Figure 2, the predicted overshoot pressure approximately equals maximum measured pressures, and the Stauffer and Dracos data closely follow the predicted metastable drainage pressure curve.

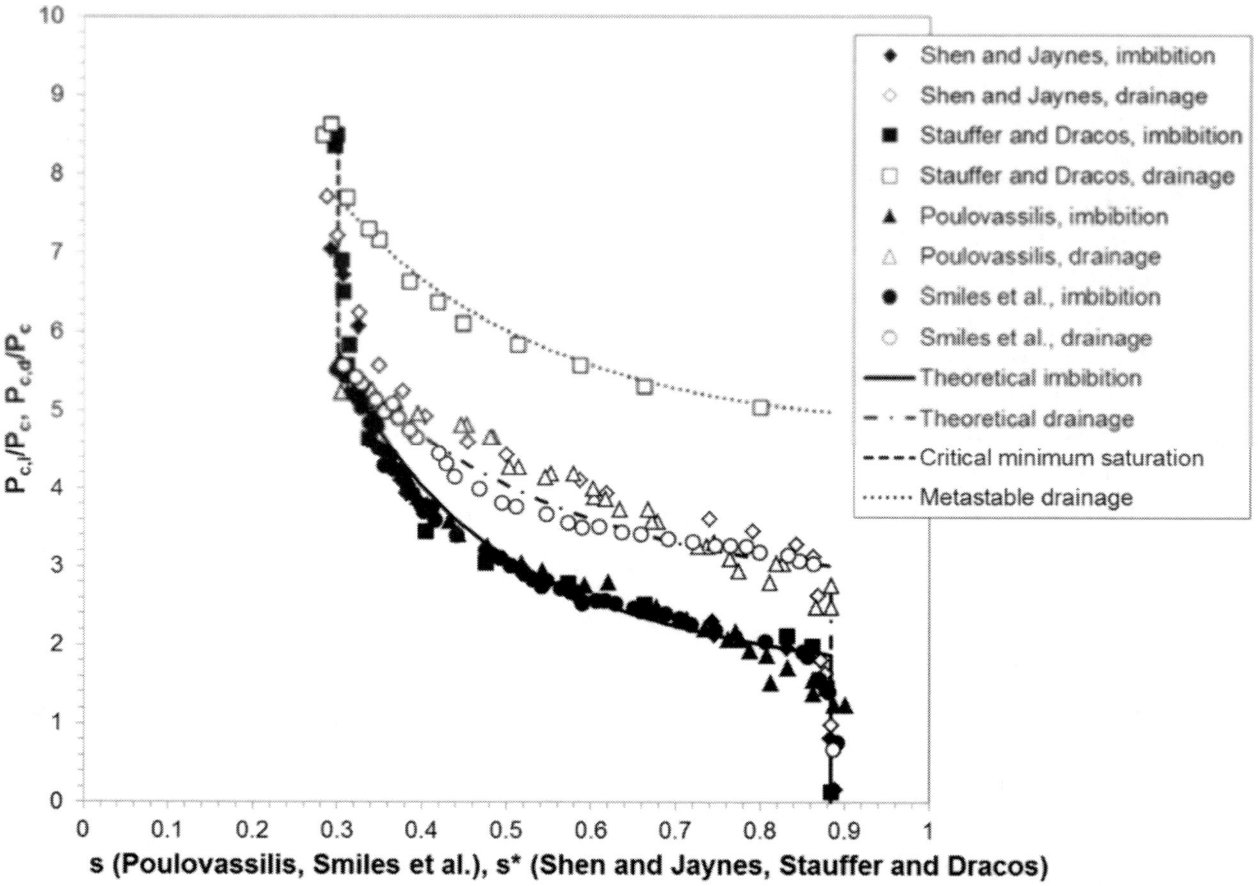

Figure 2. Variation of Imbibition and Drainage Pressures with Relative Saturation, Corrected for Fissure Effect

Evaluation of Characteristic Pore Pressures

The final step in development of the capillary pressure model is to relate the characteristic pore pressure P_c derived from the data fit to the particle size of the porous material. (One may recall that, for each data set that was regressed, the pressure multiplier is $\frac{1}{P_c}$.) From an analysis of the capillary rise in a powder, based on the specific surface area of the powder, Dunstan and White derived the following relationship between the average particle diameter and the pore diameter:[18,19]

$$d_{pore} = \frac{2\varepsilon}{3(1-\varepsilon)} \bar{d}_{part} \tag{34}$$

Average particle diameters were calculated from sieve tray data using an interval-censored method[20] based on an assumed log normal size distribution. For each data set, the void fraction was either given, calculated from the bulk density of the sand, or estimated by fitting the residual nonwetting saturation to the capillary pressure data.

The pore diameter also is related to the interfacial tension, σ, the liquid wetting angle, θ, and the capillary rise pressure, $P_{c,i,0}$, by the Laplace relation:

$$d_{pore} = \frac{4\sigma \cos(\theta)}{P_{c,i,0}} \tag{35}$$

The surface tension of water at room temperature is 72.2 dyne/cm^2.[21] The measured wetting angle is highly variable and depends on the measurement method. For air-dried sand, a dynamic measurement method angle yields a wetting angle of 45°.[22]

Table 1 compares pore diameters calculated from the average particle diameter (Equation 34) and the capillary rise pressure to the characteristic pore diameter for the capillary pressure model presented in this study (Equation 35). As may be seen, these pore diameters are in relatively close agreement; the pore diameter calculated from the capillary pressure model differs from the pore diameter calculated from the average particle diameter by an average of 7%, with a standard deviation of ±25%.

Table 1. Comparison of Pore Diameters from Particle Diameter and Capillary Pressure Measurements

Data Set	$\bar{d}_{part}$, µm	d_{pore} from $\bar{d}_{part}$, µm	d_{pore} from $P_{c,i,0}$, µm	Deviation, %
Shen and Jaynes	354	200	160	-20
Stauffer and Dracos	312	167	135	-19
Poulovassilas	537	185	219	18
Poulovassilas	551	154	205	33
Smiles et al.	264	100	122	23

Conclusions

A model has been developed that uses a characteristic pore pressure to predict residual wetting and nonwetting saturations and capillary imbibition and drainage pressures. The model is based on an energy balance that equates changes in potential surface energy to changes in pressure-volume work. A simple probabilistic distribution expresses capillary forces as a function of the relative saturation and the characteristic pore pressure. The model implicitly assumes that the porous medium is homogeneous and isotropic.

The model predicts a residual wetting saturation, s_{wr}, of 0.236 and a residual nonwetting saturation, s_{nr}, of 0.884. These predicted residual saturations are in almost exact agreement with measurements. In addition, the model predicts that there is a critical saturation, s_{cr}, of 0.301, below which the imbibition pressure exceeds the drainage pressure. A porous medium will not drain to any lower saturation, provided that the porous medium remains in contact with liquid and there is no external gas flow.

The capillary pressure model successfully correlates limiting imbibition and drainage pressures for selected laboratory tests that used uniformly packed, sieved sand, although for certain tests an adjusted saturation is required to account for fissures. For the data that was correlated, the characteristic pore diameter from the capillary pressure model approximately equals the pore diameter calculated from the average particle diameter using a capillary rise model. This agreement suggests that, for uniform packed sands with a relatively narrow particle size distribution, a capillary pressure curve can be estimated from a measured particle size distribution. The applicability of the model to soils and rocks with widely varying pore size distributions has not been demonstrated.

Acknowledgment

This work was funded by the U. S. Department of Energy Office of Environmental Management under contract number DE-AC09-08SR22470.

References

1. Brooks, R.H. and A.T. Corey (1964), "Hydraulic properties of porous media," Hydrol. Pap. 3, Civ. Eng. Dept., Colorado State University, Fort Collins, Colo.
2. van Genuchten, M.T. (1980), "A closed-form equation for predicting the hydraulic conductivity of unsaturated soils," Soil Sci. Soc. Am. J. 44, 892-898.
3. Kosugi, K. (1994), "Three-parameter lognormal distribution model for soil water retention," Water Resour. Res. 30(4), 891-901.
4. Kosugi, K. (1996), "Lognormal distribution model for unsaturated soil hydraulic properties," Water Resour. Res. 32(9), 2697-2703.
5. Nemes, A., M.G. Schaap, and F.J. Leij (1999), "The UNSODA unsaturated soil hydraulic database, version 2.0," U. S. Salinity Laboratory, U. S. Department of Agriculture, Agricultural Research Service, Riverside, California.
6. Nemes, A., M.G. Schaap, F.J. Leij, and J.H.M. Wosten (2001), "Description of the unsaturated soil hydraulic database UNSODA version 2.0," J. Hydrol. 251(3-4), 151-162.
7. Rossi, C. and J.R. Nimmo (1994), "Modeling of soil water retention from saturation to oven dryness," Water Resour. Res. 30(3), 701-708.
8. Parker, J.C. and R.J. Lenhard (1987), "A model for hysteric constitutive relations governing multiphase flow: 1. Saturation-pressure relations," Water Resour. Res. 23(12), 2187-2196.
9. White, M.D. and M. Oostrom (1996), "STOMP, subsurface transport over multiple phases: theory guide," U.S. DOE Pacific Northwest National Laboratory Report PNNL-11217, UC-2010.
10. Demond, A.H. and P.V. Roberts (1991), "Effect of interfacial forces on two-phase capillary pressure-saturation relationships," Water Resour. Res. 27(3), 423-437.
11. Morel-Seytoux, H.J. and J. Khanji (1975), "Prediction of imbibition in a horizontal column," Soil Sci. Soc. Am. Proc. 39, 613-617.
12. Morel-Seytoux, H.J., J. Khanji, and G. Vachaud (1973), "Prediction errors due to uncertainties in the measurement and extrapolation of the diffusivity function," Civil Engineering Report, Colorado State University, Fort Collins, Colorado 80523, CEP72-73, HJM48.
13. Shen, R. and D.B. Jaynes, (1988), "Effect of soil water hysteresis on simulation, infiltration, and distribution," Shuili Xuebao: J. Hydr. Eng. (China) 10, 11-20.
14. Stauffer, F. and T. Dracos (1986), "Experimental and numerical study of water and solute infiltration in layered porous media," J. Hydrol. 84(1-2), 9-34.

15. Poulovassilis, A. (1970), "Hysteresis of pore water in granular porous bodies," Soil Sci. 109(1), 5-12.
16. Smiles, D.E., G. Vachaud, and M. Vauclin (1971), "A test of the uniqueness of the soil moisture characteristic during transient, nonhysteric, flow of water in a rigid soil," Soil Sci. Soc. Am. Proc. 35, 534-539.
17. Vachaud, G. and J.-L. Thony (1971), "Hysteresis during infiltration and redistribution in a soil column at different initial water contents," Water Resour. Res. 7(1), 111-127.
18. White, L.R. (1982), "Capillary rise in powders," J. Colloid Interf. Sci. 90(2), 536-538.
19. Dunstan, D., and L.R. White (1986), "A capillary pressure method for measurement of contact angles in powders and porous media," J. Colloid Interf. Sci. 111(1), 60-64.
20. So, Y., G. Johnston, and S.H. Kim, "Analyzing Interval-Censored Data with SAS® Software," Paper 257-2010, Proceedings, SAS Global Forum 2010, Seattle, Washington, April 11-14, 2010.
21. Lide, D.R., ed. (1994), CRC Handbook of Chemistry and Physics, 75th ed., CRC Press, Boca Raton, Florida.
22. Weisbrod, N., T. McGinnis, M.L. Rockhold, M.R. Niemet, and J.S. Selker (2009), "Effective Darcy-scale contact angles in porous media imbibing solutions of various surface tensions," Water Resour. Res. 45(4), W00D39.

469001 Technetium Leaching Cementitious Materials

Abstract:

Technetium (Tc), a nuclear fission product with no stable isotopes, represents much of the long-term radioactivity associated with nuclear waste. Under oxidizing conditions, Tc persist in the environment as an anion in the +7 oxidation state, i.e., pertechnetate (TcO_4^-). In soil partitioning studies, the oxidized species displays limited retention generally associated with amphoteric soil oxides. However, Tc(VII) is subject to chemical reduction to the +4 oxidation state, which is less soluble and mobile in the environment. One disposition strategy for immobilizing ^{99}Tc containing waste is the production of cementitious materials that contain reducing agents, such as blast furnace slag (BFS), to enhance the retention of redox sensitive contaminants. At the Saltstone Disposal Facility (SDF) on the Department of Energyâ€™s Savannah River Site (SRS), low-level radioactive saltwaste is mixed with BFS, fly ash and portland cement to produce a solidified cementitious material known as Saltstone, which is deposited in a series of vaults for long-term disposal.

In the current study, contaminant mass transfer rates for ^{99}Tc and other contaminants from saltstone simulants spiked with ^{99}Tc and rhenium (Re), often used as surrogate for ^{99}Tc, and actual SDF saltstone samples were assessed using EPA Method 1315, a method for evaluating contaminant leaching from intact monolithic materials. Since redox state plays a key role in controlling ^{99}Tc solubility, the monoliths were exposed to one of three different test environments during leaching: aerobic (ambient air), anoxic (N_2), and reducing (H_2/N_2). Rhenium leaching rates (and other poorly sorbing contaminants like NO_3^- and iodine) were much higher than ^{99}Tc, which was attributed to poor initial reduction or more rapid oxidation, indicating that Re is not a suitable chemical analog for evaluating Tc partitioning under reducing conditions.

456482 Remediation of Uranium-Contaminated Groundwater By Functionalized Magnetic Mesoporous Silica Nanoparticles

Abstract:

U(VI) species display limited adsorption onto sediment minerals and synthetic sorbents in pH < 4 or pH > 8 groundwater. In this work, magnetic mesoporous silica nanoparticles (MMSNs) with magnetite nanoparticle cores were functionalized with various organic molecules using post-synthetic methods. The functionalized MMSNs were characterized using N_2 adsorption-desorption isotherms, thermogravimetric analysis (TGA), transmission electron microscopy (TEM), ^{13}C cross polarization and magic angle spinning (CPMAS) nuclear magnetic resonance (NMR) spectroscopy, and powder X-ray diffraction (XRD). The functionalized MMSNs were effective for U removal from pH 3.5 and 9.6 artificial groundwater (AGW). Functionalized MMSNs removed U from the pH 3.5 AGW by as much as 6 orders of magnitude more than unfunctionalized nanoparticles or silica and had adsorption capacities as high as 38 mg/g. They removed U from the pH 9.6 AGW as much as 4 orders of magnitude greater than silica and 2 orders of magnitude greater than the unfunctionalized nanoparticles with adsorption capacities as high as 133 mg/g. These results provide an applied solution for treating U contamination that occurs at extreme pH environments and a scientific foundation for solving critical industrial issues related to environmental stewardship and nuclear power production.

Further, after U adsorption, U L_3-edge XANES and EXAFS spectra of all functionalized MMSNs were collected using the Materials Research Collaborative Access Team (MRCAT) Sector 10-ID beamline at the Advanced Photon Source (APS) in order to understand molecular mechanisms of different U species binding with functional organic ligands. Such molecular mechanistic studies can further guide the future selection of organic ligands to develop more robust nano sorbent materials that are more efficient and cost-effective for U extraction from environmental aqueous systems.

472311 Kinetics of Simultaneous Transformation of Co-Existing Mixed Contaminant Media of Cr(VI) and TCE Using Nanoscale Zero Valent Iron (nZVI)

Abstract:

Industrialization has helped advance technology but at the cost of contaminating the natural environment. Several products and by-products of industrial processes are very toxic and hazardous to the health of terrestrial and aquatic life forms. Wastewater and other industrial effluents leach through the earth crust, adsorb on to the surface of soil particles and contaminate subsurface aquifers resulting in the pollution of ground water and many drinking water sources. Among the numerous pollutants co-existing in groundwater and subsurface aquifers, Hexavalent Chromium, Cr(VI), and Trichloroethylene, TCE, have been classified by the USEPA as known carcinogens. Thus, remediation efforts to remove/convert them to non-toxic forms are a priority. A survey of more recent research literature, shows many successful attempts using nanoscale Zero Valent Iron (nZVI), to individually reduce Cr(VI) to its benign form Cr(III) and de-halogenate TCE, but none of these studies have attempted to investigate simultaneous transformation of co-existing Cr(VI)â€"TCE mixed contaminant media and the associated synergistic or antagonistic effects. Using nZVI to remediate Cr(VI)-TCE mixtures has significant advantages including: very low to zero toxicity to the environment, abundance and presence of ZVI in the subsurface, and the ability to degrade TCE and Cr(VI) to less toxic forms.

Results of kinetic analysis using nZVI as a single transforming agent to degrade a mixed contaminant media of Cr(VI)-TCE in aqueous and soil systems will be presented for various pH and temperature values. Application to subsurface hazardous waste remediation contaminated with multiple contaminants will be discussed.

471994 Numerical Simulation and Optimization of Environmental Treatment Design Using Parallel Computational Methods: Theory and Application

Abstract:

ABSTRACT

Environmental remediation has been ongoing in the Unites States for several decades and has produced some successes. However, the more difficult sites remain challenging and resistant to efficient and effective remediation. These are precisely the environmental problems that remain as persistent fiscal and technological challenges as the success of the heuristic trial-and-error approach, which assisted in remediating the simpler sites, has reached it a limit of applicability on these cases. These tools herein enable formal, physically consistent and robust analysis of optimal remedial designs for environmental contamination problems of any level of complexity.

This paper discusses and demonstrates fast computational simulation and optimization technologies used for optimally managing environmental remediation project and programs by reducing remediation design analysis timeframes and costs. Those responsible for managing or remediating chemical or radionuclide impacted sediments, soils, surface or groundwater including designing remedial design and operation projects will benefit from tools that will help them find acceptable, efficient and effective solutions to complex and multi-faceted planning challenges at efficient cost points. A new universal optimization system is developed which has successfully been deployed on environmental and radionuclide impacted sites across a variety of computational environments including desktops and the high performance elastic Cloud-based computing platforms.

This computational advancement removes the impediment of burdensome and unpredictable runtime requirements for optimizing the design of these types of projects, which have typically consumed weeks or months of calendar time using heuristic trial-and-error or even automated serial computation approaches. These new optimization methods return results in minutes to hours for simulations and optimal solutions in hours to at most a few days for medium and larger size projects which here-to-fore required months to solve.

Computational Approach

PBMO™ is a universal optimization tool. Capabilities include optimizing linear or non-linear elliptic, parabolic and hyperbolic models and objective functions that are continuous, non-continuous and mixed integer. It is a tool system that helps users perform complex high-level tasks with amazing simplicity. Modes of operation include standard and expert modes, and it runs on computers with single and multiple CPUs, including local area grids and the Cloud. Because it is programmed as a set of modules using standard FORTRAN, it is platform independent. It has the capability for including applications with mixed operating systems and can deploy with multiple, independent or blended process models including subject matter expert (SME), data-driven (DD), Physics-based (PB) or integrated model (IM). Analysis can begin with the best result from a trial-and-error approach, or from a random starting point so efforts engaged to date by the design teams can be fully used. The optimal solution search relies on evaluations of candidate solutions, which can be entirely physics-based or response function or minorant approximated solution search (Deschaine, 2014).

This is accomplished by developing both a parallel optimal search strategy and the numerical matrix solver in the groundwater flow and transport modeling code. The parallel distributed optimization method will execute on one to several hundred (or more as needed) CPUs while the parallel numerical solver that has been added to the subsurface flow and transport simulator MODFLOW-SURFACT™ has demonstrated speedups as high as a factor of 9.8. Hence, the integrated parallel simulation and optimization method uses the breadth and depth search approach; iteratively producing sets of tens or 100s of candidate investigation points that are evaluated concurrently instead of the previous and widely used sequential approach (i.e. one model evaluation at a time). This enables 80-100% globally optimal solutions are regularly found expending only 5-10% of the common computational effort using this new parallel optimization method (reducing computation effort by 10 to 20 times and calendar time by a factor of 100 or more).

Additionally, practitioners conducting conceptual site model development, model calibration by auto-calibration methods and remedial design by heuristics can enjoy solution speedups by using the parallel numerical matrix solver in the physics-based simulator as well as automated uncertainty analysis using the Latin Hypercube, Monte-Carlo or geostatistical realization techniques. Combining investigation and design analysis with breadth and depth parallel optimization methods, projects at the fringe of practical optimization, along with projects that here-to-fore were beyond the reach of formal optimization methods are now wholly feasible and tractable optimization problems.

Remedial Design Approach

PBMO™ project design involves selecting the appropriate combination of its fully comprehensive solution options, including the full capability of mixing and matching various approaches.

- Select proper descriptive model or model set that adequately describes the system to be optimized
- Select optimization algorithm or set of proper optimization algorithms to optimize the system best.
- Select computer operation system or a mixture of operating systems, to enable the integrated system to run as-is.

It is important to note that “optimization” refers not just to cost minimization but also to the effective and efficient balance of cost, performance, risk, management, and societal priorities along with uncertainty analysis. To wit, the PBMO™ process formally integrates the stakeholder input into the analysis and provides not only the "optimal solution" but also "what-if" capability that includes management override control on remedial design analysis. This approach integrates all of these elements into a single decision framework that is tractable, traceable, and defensible. The approach is modular and scalable. It can be applied either as individual components or in total. This solution methodology represents a significant improvement over the non-optimal "trial and error" approach to developing environmental response(s).

METHODS

Understanding optimization rationale and theory, methodological strength, applicability, and limitations, is critical for understanding why optimization is valuable and how to successfully and deploy computational optimization analysis on projects. Project implementation experience has shown that several effective optimization methods exist that are routinely successful when used on projects and are included in this system. Exhibition of case histories demonstrates their value for developing optimal solutions not obtainable by traditional subjective design methods (a.k.a. manual trial and error).

PBMO™ uses a breadth and depth search strategy. The breadth search identifies good solution quickly, whereas the local search drills down to find the optimal value. The optimization methods use include:

- Derivative-based global methods
  - Linear programming (LP)
  - Sequential linear approximation (SLA)
  - Sequential Quadratic Programming (SQP)
  - Generalized Reduced Gradient (GRG)
- Derivative-free global methods
  - Adaptive Design of Experiment (ADOE)
  - Adaptive Global Random Search (AGRS)
  - Branch and Bound (B&B)
  - Global Adaptive Random Search (GARS)

Table 1 depicts the applicability of the PBMO™ system (Deschaine, 2014).

Table 1. PBMO™ Solution Scope and Analysis Results

Scope	Description	Affects	Analysis Results
Optimization of environmental *response program*	Optimal FS/CMS planning-level design considering all reasonably viable alternative remediation responses including waste isolation and monitoring.	Total response cost, multi-period fiscal budget planning and allocation strategies; calculations and contingencies	Optimal and balanced solution that best meets needs of site and stakeholders
Optimization of environmental *remedial*	Optimal detailed design of FS/CMS selected	Total response cost, specific capital	Optimal design basis for site including

design	response program	expenditures, annual costs, plant operations	constructability assessment and predicted performance
Optimization of *long-term monitoring*	Optimal design of monitoring well network and sampling program	Total response cost, annual costs, stakeholder confidence that the remedial design and installation are operating as intended.	Proactive management of stakeholder expectations, early warning detection if as-built and operated systems' performance begins to deviate from design intent.
Optimal *contingency plans*	Optimal and pro-active considerations for addressing remedial design *and observed* environmental system response	Total response cost, annual costs, stakeholder confidence that the remedial program and design are robust and amenable to refinement, if warranted.	Optimal and robust remedial system designs, installations and monitoring that incorporate feedback from the operation. Efficiently allow for modifications or inclusion of additional components or processes.

The solution from the PBMO™ methodology has always exceeded the subjective engineering solution when available for comparison (Deschaine, 2014).

CONCLUSIONS

The basic PBMO™ methodology has been previously tested, verified, validated and documented using a broad range of industrial projects. Those results indicate that it meets or exceeds the best available solution either in use (i.e. as an RIP), proposed by the subject matter experts or developed by other optimization methods. These two new parallelization methods now allow for the optimization of remedial designs for the more complicated problems. These are precisely the environmental problems that remain in the environmental contamination site portfolio as persistent challenges. The earlier successes of the heuristic trial-and-error approach - which assisted in remediating the simpler sites - has reached it limit of applicability in these cases. By simultaneously combining project level optimization with optimal programmatic design, the optimal design and implementation of environmental challenge response are now computationally tractable, practical and viable.

REFERENCES

Deschaine, L. M. (2014). Decision support for complex planning challenges. Ph.D. Dissertation, Chalmers University of Technology, Göteborg, Sweden. 233p.

Deschaine, L. M., Lillys, T. P., & Pintér, J. D. (2013). Groundwater remediation design using physics-based flow, transport, and optimization technologies. Environmental Systems Research, 2(1), 1-21.

http://www.environmentalsystemsresearch.com/content/pdf/2193-2697-2-6.pdf

THE REACTION RATE OF URANIUM-ALUMINUM ALLOYS IN BOILING CAUSTIC SOLUTIONS*

Fant, Brian R., Huxtable, William P., and Swientoniewski, Mark D.,

Bechtel National, Inc., Oak Ridge, Tennessee

Abstract

This effort correlates and compares historical data on dissolution rates of aluminum (Al) and uranium-aluminum (U-Al) alloys in boiling caustic (NaOH) solutions to support a dissolution reactor design. First, a new reaction rate model (the Modified Wymer model) is proposed to correlate dissolution rate data for Al and U-Al alloys in NaOH. Second, bounding metal dissolution rates and hydrogen production rates are proposed for Al and U-Al alloys in boiling 6 M NaOH solutions. Third, an inconsistency in Al reaction rates in 6 M NaOH by Wymer et al.[1] is addressed. These dissolution rates are first-order reactions as a function of dissolvent NaOH concentration and are correlated for various Al and U-Al alloy compositions as well as the presence of sodium nitrate ($NaNO_3$) in the dissolvent. Bounding dissolution rates and hydrogen production rates are predicted based on a maximum 6 M NaOH concentration expected in the dissolvent.

Introduction

This study has three objectives:

1. To propose a rate model to correlate dissolution rate data for Al metal and U-Al alloys in boiling NaOH dissolvent.
2. To propose bounding dissolution rates and hydrogen production rates for Al and U-Al alloys in boiling 6 M NaOH solution.

3. To address an inconsistency by Wymer et al.[1] in a reported Al dissolution rate that impacts the bounding dissolution rate estimate.

The scope of this study is to collect and correlate both contemporary and historical data on dissolution rates of Al metal and U-Al alloys in boiling NaOH solutions. These dissolution rates are modeled as a first-order reaction of Al metal or U-Al alloys with respect to boiling NaOH either with or without $NaNO_3$ in the dissolvent (for the purpose of suppressing hydrogen gas). Bounding dissolution rates and hydrogen production rates are predicted based on a maximum 6 M NaOH concentration in the dissolvent.

All aluminum metal alloys used in the referenced sources were different types of commercially pure aluminum (99+% Al) alloys. As is shown in this study, the effect of the small amounts of alloying components in these commercially pure aluminum alloys can dramatically reduce dissolution rates in caustic.

Chemistry of U-Al Alloy Dissolution in Caustic Solutions

Wymer et al.[1] report that when U-Al slugs are dissolved in sodium hydroxide solution, the primary reaction is:

$$Al + NaOH + H_2O \rightarrow NaAlO_2 + 3/2H_2 \qquad \text{Equation 1}$$

A secondary, simultaneous reaction results in the formation of solid uranium oxides, with its reaction rate being inconsequential. Woerner and Taylor[2] observed that Aluminum Alloy 1100 (Alloy 1100) starts to rapidly react with 1 M to 3 M NaOH dissolvent at about 60°C to 65°C, and this exothermic reaction causes the dissolvent to reach the boiling temperature in a few seconds. Based on Equation 1, the maximum hydrogen production per kilogram of aluminum dissolved can be estimated as:

$$Q_{H_2} = V_{std} \times \frac{3}{2}\frac{moles\ H_2}{mole\ Al\ dissolved} \times \frac{1000\ gm/kg}{AW_{Al}} = \frac{1246\ L\ H_2}{kg\ Al} \qquad \text{Equation 2}$$

Where,

V_{std} is the volume of 1 mole of gas at standard conditions = 22.414 L/mole from Mohr[3]

AW_{Al} is the atomic weight of aluminum = 26.98 gm/mole from Wieser et al.[4]

From Snyder and Davis[5] for a 6 M NaOH dissolvent, if there are at least 50 grams of $NaNO_3$ per liter (0.6 M $NaNO_3$) in the NaOH dissolvent, the hydrogen generation rate decreases from 1246 ml/gm to only 2 to 3 ml/gm of Al dissolved. This reduction in reaction rate is likely due to the more complex global reaction depicted in Equation 3. Instead of hydrogen, a lesser volume of ammonia is produced, which is refluxed in most experiments to maintain dissolvent volume.

$$8Al + 5NaOH + 3NaNO_3 + 2H_2O \rightarrow 8NaAlO_2 + 3NH_3 \qquad \text{Equation 3}$$

The molar ratio (3:8) of $NaNO_3$ to Al in this stoichiometry is sufficient to react all the H_2 formed in Equation 1 to produce ammonia. As this ratio is increased to 3:2, Gresky[6] reports that sodium nitrite is increasingly produced instead of ammonia:

$$2Al + 2NaOH + 3NaNO_3 \rightarrow 2NaAlO_2 + 3NaNO_2 + H_2O \qquad \text{Equation 4}$$

This reaction consumes more reactants and produces more product waste per mole of Al dissolved. As shown in Figures 3.1-1 and 3.1-2 in Gresky,[6] the fractions of Al dissolved by Equation 3 and Equation 4 as a function of excess $NaNO_3$ can be calculated. A molar excess of 20% NaOH is recommended by Gresky[6] to minimize waste and suppress hydrogen production based on the stoichiometry of Equation 3.

Unfortunately, no conclusive stoichiometry for uranium dissolution in NaOH/$NaNO_3$ could be identified in the literature. Charpie et al.[7] report that only uranium aluminate solids (i.e., mainly UAl_4 and some UAl_3) are produced by NaOH dissolution of U-Al alloys without $NaNO_3$. For NaOH/$NaNO_3$ dissolution of U-Al alloys, Charpie et al.[7] report that all uranium is present as sodium diuranate solids ($Na_2U_2O_7$). Following Gresky's approach[6] and given the observed $Na_2U_2O_7$ product with the NaOH and $NaNO_3$ reactants, the most likely products of NaOH/$NaNO_3$ dissolution of U in U-Al alloy include NH_3, H_2, and $NaNO_2$. Cunningham and Taylor[8] observed that only ammonia gas and water vapor were present in the offgas of their experiments. Based on these observations, there are two competing reaction stoichiometries:

$$4U + NaOH + 3NaNO_3 + 4H_2O \rightarrow 2Na_2U_2O_7 + 3NH_3 \qquad \text{Equation 5}$$

$$4U + 4NaOH + 12NaNO_3 \rightarrow 2Na_2U_2O_7 + 12NaNO_2 + 2H_2O \qquad \text{Equation 6}$$

Unlike for aluminum dissolution in NaOH/$NaNO_3$, there are no data to determine the product split between Equation 5 and Equation 6 for uranium dissolution. To bound waste production, the stoichiometry in Equation 6 produces the maximum amount of waste per mole of U dissolved. Alternatively, the same percentage split as for aluminum dissolution could be used for uranium dissolution. If most of the expected U-Al alloy to be processed contains 8% U (compositions of U in U-Al Alloys are expressed as weight percent), the mole fraction of U is near 0.01, so either approach gives mass balance results with less than 1% error. Even for a

U-Al alloy containing 55% U, the mole fraction of uranium is low (0.12). This supports the assertion that the uranium reaction is secondary and inconsequential.

Resolution of a Dissolution Rate Discrepancy

Wymer and Blanco[9] presented the following rate equation to fit their data on the dissolution of aluminum in 7.5% U in U-Al alloy slugs in boiling NaOH solution:

$$-\frac{dC}{dt} = k * \frac{s}{V} * C \qquad \text{Equation 7}$$

Where,

k is the observed specific reaction rate constant, ~2.6*10^{-4} L/cm^2-min
s is the surface area of the alloy shape dissolved, cm^2
V is the solution volume used in the dissolution, assuming total reflux of any condensate generated by the dissolution reaction, L
C is the molar concentration of sodium hydroxide, moles/L
t is the time, minutes

The observed specific reaction rate constant in Equation 7 is stated by Wymer et al.[1] without reaction rate data to substantiate it. Snyder and Davis[5] recommend 5.5 M NaOH as the optimum solution for dissolving U-Al alloys in NaOH/$NaNO_3$ with minimum waste and hydrogen production. Wymer et al.[1] state that the observed specific reaction rate constant "is equivalent to an aluminum dissolution rate of 36 mg Al/cm^2-min in 6 M NaOH." However, this stated dissolution rate is not consistent with that calculated using Equation 7. The dissolution rate in NaOH can be correctly calculated as shown in Equation 8.

Equation 7 can be rearranged to calculate the reaction rate in units of mg Al/cm^2-min as follows:

$$\frac{1}{s}\frac{dm_{Al}}{dt} = k * C * A * AW_{Al} * 1000\frac{mg}{gm} \qquad \text{Equation 8}$$

Where,

m_{Al} is the mass, milligrams
A is the moles of Al dissolved per mole of NaOH consumed = 1 for NaOH dissolvent (Equation 1)

By substitution into Equation 8, the calculated reaction rate in boiling 6 M NaOH dissolvent is the following:

$$\frac{1}{s}\frac{dm_{Al}}{dt} = \frac{2.6 * 10^{-4} L}{cm^2\, min} * \frac{1\, mole\ Al}{mole\ NaOH} * \frac{6\, mole\ NaOH}{L} * \frac{26.98\, gm\ Al}{mole\ Al} * \frac{1000\, mg}{gm} = 42\frac{mg\ Al}{cm^2 min}$$

For process safety, this estimate of the maximum aluminum dissolution rate results in a more conservative value for estimating hydrogen production and is preferred. The following sections show that a 42 mg Al/cm^2-min maximum Al dissolution rate is bounding for the dissolution of U-Al alloys containing at least 5% U in boiling NaOH or NaOH/$NaNO_3$ containing up to 6 M NaOH.

Dissolution Rates of Aluminum and U-Al Alloys in Caustic Solutions

To correlate the available dissolution rate data, an intercept was determined to be needed in the Equation 8 rate model provided by Wymer et al.[1] as follows:

$$-\frac{dC}{dt} = \frac{s}{V} * (k * C + r_0) \qquad \text{Equation 9}$$

Where,

r_0 is the y-intercept from the linear fits of aluminum reaction rate data given in Figure 1.

This new, caustic dissolution reaction rate model is referred to as the Modified Wymer model. Solving for the reaction rate, *r*, in mg/cm^2-min, as before, gives the following reaction rate model for correlating each set of the aluminum reaction rate data in Figure 1:

$$r = \frac{1}{s}\frac{dm_{Al}}{dt} = \left(k * A * AW_{Al} * 1000\frac{mg}{gm}\right) C + r_0 \qquad \text{Equation 10}$$

The slope for each linear fit in Figure 1 is the product of the constants given in Equation 8:

$$Slope_{NaOH} = k * A * AW_{Al} * 1000\frac{mg}{gm} = 26980k \qquad \text{Equation 11}$$

$$Slope_{NaOH/NaNO_3} = k * A * AW_{Al} * 1000\frac{mg}{gm} = 43168k \qquad \text{Equation 12}$$

The specific reaction rate constant, *k*, has a unique value for each material composition and depends on whether $NaNO_3$ is used to suppress hydrogen production.

To ensure the bounding aluminum reaction rates are established, historical NaOH dissolution rates for Al and U-Al alloys are compared and correlated using Equation 10. This comparison reveals a basis for bounding Al reaction rates for Al alloys and U-Al alloys with less than 8% U. The correlation of historical reaction rate data as a first-order linear reaction with NaOH and NaOH/$NaNO_3$ solutions provides practical dissolution rate models.

Figure 1 compares historical NaOH dissolution data for Al or U-Al alloy materials from Wymer et al.,[1] Snyder and Davis,[5] Gresky,[6] Woerner,[10] and Woerner and Patton.[11] Average dissolution rates for the dissolution of 5% U in U-Al alloy in $NaOH/NaNO_3$ reported by Burns et al.[12] are not included in Figure 1. The Burns et al. data[12] cannot provide penetration rates (linear distance inward of mass removed from the surface) as a function of NaOH concentration without making several assumptions including an assumed first-order linear reaction, which is clearly demonstrated in Figure 1.

The dissolution rates depicted in Figure 1 are each reported in Tables 1 through 6 and converted to mg/cm^2-min, as necessary. The graph of the Table 1 data for Wymer et al.[1] is calculated using Equation 8. In Table 2, the NaOH concentration is the initial concentration (5.5 M) minus 5/8 times the Al molarity based on Equation 3. In Table 3, penetration rates are interpolated from Figure 3.1-3 of Gresky[6] and converted by multiplying by the product of the density of Al (2.7 mg/mm^3 from Lide[13]) and a conversion factor [(100 mm^2/cm$^{2)}$/(60 min/hr)] to obtain the reaction rate in mg Al/cm^2-min. In Table 4, the Woerner[10] molar rate is converted to mg Al/cm2-min by multiplying by the product of the molecular weight of aluminum (AW_{Al}) and a conversion factor [(1000 mg/gm)/(60 min/hr)]. In Tables 5 and 6, reaction rates are calculated based on data presented by Woerner and Patton,[11] and Figure 3.1-3 of Gresky[6] is used to calculate the NaOH concentration from the measured aluminum dissolved.

There are four general observations related to Figure 1:

1. Only dissolution rate data above 1.5 M NaOH are included or were available.
2. The dissolution rate of aluminum and various aluminum-uranium alloys is first order (linear) with respect to NaOH molarity.
3. The specific reaction rate constant (slope) is a function of aluminum purity and alloy composition.
4. $NaNO_3$ reduces the dissolution rate of aluminum and U-Al alloys.

Regarding the first observation, Snyder and Davis[5] recommend a minimum of 1.5 M NaOH in the dissolvent to prevent the precipitation of hydrated alumina ($Al_2O_3 \bullet H_2O$) that would contaminate the solid $Na_2U_2O_7$ product. Wymer and Blanco[9] recommend an even greater concentration, 3.0 M NaOH, in the dissolvent for long term stability. Therefore, in Figure 1, only reaction data for greater than 1.5 M NaOH are of practical interest.

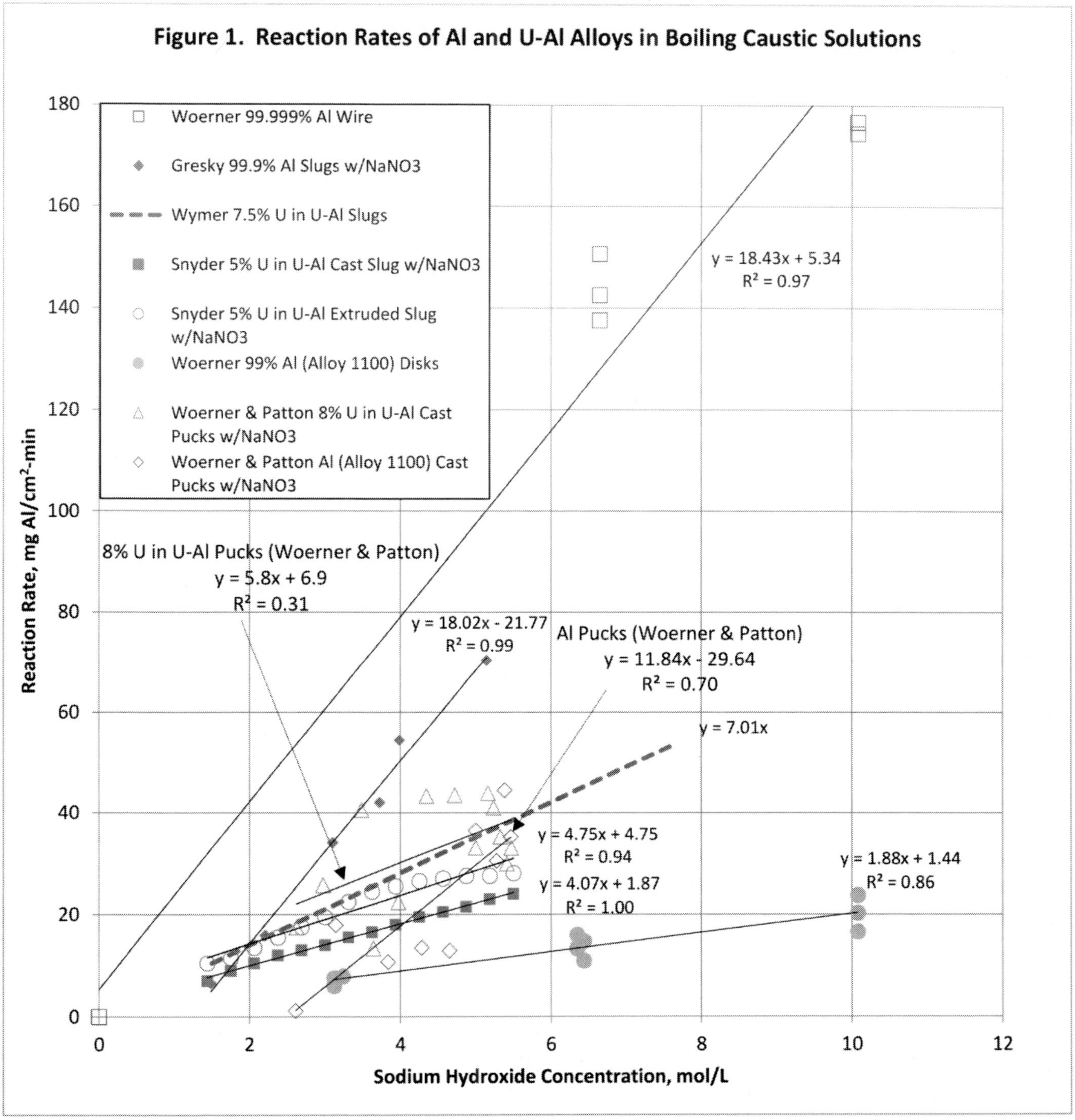
Figure 1. Reaction Rates of Al and U-Al Alloys in Boiling Caustic Solutions
Woerner 99.999% Al Wire
Gresky 99.9% Al Slugs w/NaNO3
Wymer 7.5% U in U-Al Slugs
Snyder 5% U in U-Al Cast Slug w/NaNO3
Snyder 5% U in U-Al Extruded Slug w/NaNO3
Woerner 99% Al (Alloy 1100) Disks
Woerner & Patton 8% U in U-Al Cast Pucks w/NaNO3
Woerner & Patton Al (Alloy 1100) Cast Pucks w/NaNO3
y = 18.43x + 5.34
R² = 0.97
8% U in U-Al Pucks (Woerner & Patton)
y = 5.8x + 6.9
R² = 0.31
y = 18.02x - 21.77
R² = 0.99
Al Pucks (Woerner & Patton)
y = 11.84x - 29.64
R² = 0.70
y = 7.01x
y = 4.75x + 4.75
R² = 0.94
y = 4.07x + 1.87
R² = 1.00
y = 1.88x + 1.44
R² = 0.86
Reaction Rate, mg Al/cm²-min
Sodium Hydroxide Concentration, mol/L
0
20
40
60
80
100
120
140
160
180
0
2
4
6
8
10
12

Table 1. Calculated Aluminum Reaction Rate based on Wymer et al.[1]

NaOH, mole/L	7.5% U-Al Slugs in NaOH, mg Al/cm^2-min
1.50	11
2.00	14
2.50	18
3.00	21
3.50	25
4.00	28
4.50	32
5.00	35
5.50	39
6.00	42
6.50	46
7.00	49
7.50	53
7.65	54

Table 2. Reaction Rate Data from Snyder and Davis[5]

Aluminum Conc., mole/L	NaOH Conc., mole/L	5% U-Al Cast w/$NaNO_3$, mg Al/cm^2-min	5% U-Al Extruded w/$NaNO_3$, mg Al/cm^2-min
6.50	1.44	7.0	10.5
6.00	1.75	9.0	11.5
5.50	2.06	10.5	13.5
5.00	2.38	12.0	15.5
4.50	2.69	13.0	17.5
4.00	3.00	14.0	19.5
3.50	3.31	15.5	22.5
3.00	3.63	16.5	24.5
2.50	3.94	18.0	25.5
2.00	4.25	19.5	26.5
1.50	4.56	20.5	27.0
1.00	4.88	21.5	27.5
0.50	5.19	23.0	27.5
0.00	5.50	24.0	28.0

Table 3. Aluminum Reaction Rate Data from Gresky[6]

NaOH Conc., mole/L	Penetration Rate, mm/hr	99.9% Pure Al Slugs, mg Al/cm^2-min
1.48	1.47	7
2.21	3.54	16
3.10	7.56	34
3.73	9.35	42
3.99	12.10	54
5.15	15.65	70

Table 4. Reaction Rate Data from Woerner[10]

NaOH Molarity	Rate, mole Al/cm^2-hr	99% Pure Al Alloy 1100 Disks, mg Al/cm^2-min	Ultrapure 99.999% Al Wire, mg Al/cm^2-min
3.12	0.0168	7.6	
3.12	0.0134	6.0	
3.24	0.0176	7.9	
6.44	0.0328	14.7	
6.44	0.0243	10.9	
6.35	0.0296	13.3	
6.35	0.0355	16.0	
6.35	0.0295	13.3	
10.08	0.0368	16.5	
10.08	0.0527	23.7	
10.08	0.0451	20.3	
0	0.0000		0.0
0	0.0000		0.0
0	0.0000		0.0
3.33	0.0069		3.1
3.33	0.0075		3.4
3.33	0.0091		4.1
6.65	0.3350		150.6
6.65	0.3060		137.6
6.65	0.3170		142.5
10.08	0.3880		174.5
10.08	0.3930		176.7

Table 5. Woerner & Patton[11] 8% U in U-Al Cast Pucks w/$NaNO_3$

Start Time (min)	Final Time (min)	Avg. Al Dissolved (M)	Avg. U Dissolved (M)	Avg. NaOH Concentration (M)	Reaction Rate (mg/min-cm^2)
0	0.5	0.04	0.00	5.47	32.93
0.5	1	0.11	0.00	5.41	29.86
1	2	0.22	0.00	5.33	35.14
2	2.33	0.33	0.00	5.24	41.00
2.33	3	0.42	0.00	5.16	43.89
3	5	0.63	0.01	5.00	32.99
5	7.5	0.98	0.01	4.72	43.49
7.5	10	1.45	0.01	4.34	43.34
10	15	1.92	0.02	3.97	22.42
15	20	2.33	0.02	3.64	13.37
20	24	2.53	0.03	3.48	40.51
24	35	3.17	0.03	2.97	25.76
35	42.75	3.62	0.04	2.62	17.48

Table 6. Woerner & Patton[11] Al (Alloy 1100) Cast Pucks w/$NaNO_3$

Start Time (min)	Final Time (min)	Avg. Al Dissolved (M)	Avg. NaOH Concentration (M)	Reaction Rate (mg/min-cm^2)
0	0.5	0.05	5.46	35.15
0.5	1	0.15	5.38	44.43
1	2	0.28	5.28	30.40
2	5	0.64	5.00	36.46
5	10	1.08	4.65	12.91
10	20	1.55	4.28	13.48
20	30	2.11	3.84	10.71
30	64	3.00	3.14	17.92
64	78	3.68	2.61	1.14

In Figure 1, the resulting correlation of Woerner's[10] aluminum wire and disk dissolution rates in NaOH is useful for comparison but not useful for dissolver design. First, Woerner[10] did not keep aluminum wire or disk samples submerged as recommended by Wymer et al.,[1] so these data cannot be compared to other historical results for predicting reaction rates of submerged Al metal. Second, unlike the Al Alloy 1100 disks dissolved by Woerner,[10] the surface area of ultrapure wire is difficult to predict as dissolution proceeds, and wire is not a representative shape for studying surface-area-dependent reaction rates of bulk metal. Finally, Woerner[10] observed that ultrapure aluminum wire reaction rates were inconsistently low in 3.33 M NaOH. Woerner[10] cut short one of his three wire dissolution experiments. Since these three experimental points have almost identically slow reaction rates, these data were excluded from Figure 1 and are not discussed further. For continuity and consistency with Gresky's data,[6] three points were added at the origin of Figure 1 and Table 4 for the linear correlation of Woerner's wire dissolution rate data.[10]

Aluminum disk and puck dissolution experiments by Woerner[10] and Woerner and Patton[11] used Alloy 1100, which is 99% Al alloyed with 0.02% to 0.05% copper, 0.10% zinc, and up to 0.95% iron and silicon.[14] Figure 1 shows that corrosion-resistant aluminum Alloy 1100 discs dissolved much slower than 5-8% U in U-Al alloys, but when the same material was fully submerged as pucks, the initial dissolution rate was about the same as for 5-8% U in U-Al alloy slugs and pucks in dissolvent with 5.5 M NaOH and 3.7 M $NaNO_3$. As aluminum slugs or pucks dissolve, the influence of alloying materials and the accumulation of their oxides, particularly iron oxide, on the surface of dissolving Al are conjectured to reduce the Al dissolution rate. This can explain the difference in the dissolution rates of Al Alloy 1100 pucks by Woerner and Patton[11] compared to 99.9% Al slugs by Gresky[6] in $NaOH/NaNO_3$, as shown in Figure 1. Based on Figure 1, the presence of uranium and the buildup of $Na_2U_2O_7$ solids apparently do not affect the dissolution rate of U-Al alloys as much as the alloying materials used in Al alloys.

From Figure 1, the calculated reaction rate is reasonably linear with respect to caustic concentration for Al pucks for Woerner and Patton's[11] data. However, the calculated reaction rate data for dissolving 8% U in U-Al pucks in $NaOH/NaNO_3$ give a much lower regression correlation, R^2, of 0.31. The data scatter is too great to rely on these data for process design, but the regression is essentially identical to Wymer and Blanco's correlation in Equation 7.[9]

The data scatter in the 8% U in U-Al puck dissolution data by Woerner and Patton[11] is not readily explainable. The initial reaction rates at 5.5 M NaOH were suppressed because the pucks were at room temperature when added to the boiling NaOH dissolvent. Unlike other studies, offgases were not condensed and water and ammonia were not returned to the reaction vessel. The loss of offgas water vapor and ammonia was compensated by manually replenishing the dissolution beaker with room temperature distilled water. As a result, the concentration of condensed ammonia returned to the beaker was negligible, the concentration of caustic was an estimate (and varied inconsistently based on the replenishing of water), and the reaction rate was periodically retarded by the addition of cooler than boiling water. These inconsistencies and differences likely account for the scatter in reaction rates that are calculated based on Woerner and Patton's[11] data. It is hypothesized that greater dissolution rates and shorter total dissolution times may be achieved by not returning condensed offgas, but the data scatter suggests this approach may not work reliably and these data are insufficient to form the basis of a reactor design.

With the exception of Woerner and Patton's[11] 8% U in U-Al puck dissolution data, Figure 1 shows that the dissolution rates of aluminum and U-Al alloys tested in boiling NaOH are well represented as first order with respect to the molar concentration of NaOH. The Snyder and Davis[5] reaction rates of bare, extruded material are the least linear of the historical data, but the regression correlation, R^2, explains almost 94% of the variation. Also, the U-Al alloy reaction rates are bounded by the more rapid rates reported for the dissolution of 7.5% U-Al in NaOH by a correlation of Wymer et al.,[1, 9] but no data are presented to perform a regression correlation. The contemporary reaction rates for Al and 8% U in U-Al pucks calculated from Woerner and Patton's data[11] correlate more poorly with the estimated NaOH concentration.

Table 7 summarizes the regression constants and specific reaction rate constants for all the aluminum and U-Al alloys tested. Based on the linear regressions given in Figure 1 and a note by Wymer et al.,[1] the reaction rate correlations should be applicable from 1.5 M NaOH to at least 7.65 M NaOH. Table 7 also compares the aluminum reaction rates and hydrogen production rates of the tests for aluminum and U-Al alloys in boiling 6 M NaOH with and without $NaNO_3$.

Unalloyed aluminum reacts much more vigorously with NaOH than U-Al alloys. From Table 7, the reaction rate of NaOH with pure Al without $NaNO_3$ present (151 mg/cm^2-min) is 3.6 times faster than for 7.5% U in U-Al alloy (42 mg/cm^2-min). A lack of solid products when dissolving pure Al may be a factor in this difference. The precise purity of the Al dissolved by Gresky[6] at Oak Ridge National Laboratory (ORNL) is not known. However, at the time, ORNL was using 99.9% Al to make 2% U, 5% U, and 7.5% U in U-Al alloys according to Hamby,[15] and the Al purity in the ORNL U-Al alloy produced was 99.75% according to Burns et al.[12] The reduced reaction rates in $NaOH/NaNO_3$ observed by Gresky[6] at ORNL are consistent with dissolution rates for Al in NaOH by Woerner.[10]

Wymer et al.[1] indicated that the Al reaction rates in NaOH for 5% U in U-Al alloy and 7.5% U in U-Al alloy are essentially the same. Based on Table 7, the addition of $NaNO_3$ to NaOH reduces the Al dissolution rate of pure Al from 151 mg/cm^2-min to 86 mg/cm^2-min, or 43%. In $NaOH/NaNO_3$ dissolvent, increasing the U in U-Al alloy slugs from 0% to 5% reduces the dissolution rate from 86 mg/cm^2-min to 26 mg/cm^2-min, or 70%. From Figure 1, the Al dissolution rates for 7.5% U in U-Al alloy in NaOH alone by Wymer et al.[1] bound the dissolution rates of 5% U in U-Al alloy extruded material and slugs in $NaOH/NaNO_3$. Higher concentrations of U in U-Al alloys should result in reduced dissolution rates than the dissolution rates from Wymer et al.,[1] as discussed in the next section. Therefore, Wymer and Blanco's correlation,[9] Equation 7, should bound the maximum Al reaction rate for U-Al alloys containing 5% or more

U. Snyder and Davis[5] observed that extruded U-Al alloy dissolves at about the same rate as cast slugs.

Table 7. Specific Reaction Rate Constants for NaOH Dissolution of Al and U-Al Alloys

Material Dissolved	Linear Fit Intercept on Fig. 1	Linear Fit Slope on Fig. 1	k, Specific Reaction Rate Constant, L/cm^2-min x 10^4		Al Dissolution Rate in 6 M NaOH, mg/cm^2-min	H_2 Production Rate in 6 M NaOH, std. ml/cm^2-min
			NaOH	NaOH/ $NaNO_3$		
99.999% Al Wire	5.34	18.43	6.83		151	188
99.9% Al Slugs	-21.77	18.02		4.17	86	0.26
7.5% U U-Al Alloy Slugs	0	7.01	2.60		42	52
Al Alloy 1100 Disks	1.44	1.88	0.70		13	16
5% U-Al Cast	1.87	4.07		0.94	26	0.079
5% U-Al Extruded	4.75	4.75		1.10	33	0.10
8% U U-Al Cast Pucks	6.9	5.8		1.34	42	0.13
Al Alloy 1100 Pucks	-29.6	11.8		2.74	41	0.12

Note: Negative or non-zero intercepts are the result of extrapolating linear fits of reaction rate data that are correlated above 1.5 M NaOH to 0 M NaOH. In reality, if the NaOH concentrations approach 0 M, reaction rates will also approach zero.

Except for ultrapure aluminum, the Al dissolution rates reported in Table 7 are for boiling 6 M NaOH using the calculated specific reaction rate constant. For ultrapure aluminum wire, the greatest of the three measured Al reaction rates of 151 mg/cm^2-min at approximately 6.65 M NaOH reported by Woerner[10] is provided in Table 7 for conservatism. The H_2 production rates in Table 7 for materials dissolved in NaOH without the presence of $NaNO_3$ are based on the stoichiometry given in Equation 1. For materials dissolved in NaOH containing at least 50 gm/L $NaNO_3$, a hydrogen production rate of 3 ml H_2 per gm of Al dissolved, as reported by Gresky,[6] is assumed in Table 7. This hydrogen production rate is based on an Al/NaOH/$NaNO_3$ molar ratio of 8/5/3 in the dissolvent (i.e., the stoichiometry in Equation 3) and excess $NaNO_3$ and NaOH based on Figure 3.1.5 in Gresky.[6]

Dissolution Rates of U-Al Alloys with Greater Than 8% Uranium

Uranium-aluminum alloys with greater than 8% uranium may need to be processed. No quantitative NaOH dissolution rate data could be found for U-Al alloys with a uranium content greater than 8%. The dissolution rate of U-Al alloys containing more than 8% U in NaOH/$NaNO_3$ dissolvent is not established nor is the rate of hydrogen production.

From Table 1 at 6 M NaOH, adding 5% U to pure Al reduced the Al dissolution rate by 70%. Therefore, slower reaction rates are expected as the U concentration is further increased in U-Al alloys. However, at the limit, pure uranium will not dissolve in $NaOH/NaNO_3$. Gresky[6] dissolved Al jackets on uranium metal slugs in $NaOH/NaNO_3$ dissolvent. Although the reaction rate was not quantified, Cunningham and Taylor[8] observed that a U-Al alloy with 10% to 13% U reacted fast enough to make a dissolvent of 3 M NaOH and 3.5 M sodium nitrate heat rapidly from an initial temperature of 60°C to boiling. Although not quantified, it appears that a U-Al alloy with up to 13% U can be as easily dissolved as a U-Al alloy with 5% to 8% U. This result and the rate data in Figure 1 support Wymer and Blanco's assertion that the dissolution rates of U-Al alloys containing between 5% and 15% U are indistinguishable.[9]

For a U-Al alloy with more than 13% U, insight into the expected reaction rate in $NaOH/NaNO_3$ may be gained from examining a U-Al phase diagram and photomicrographs of Jones et al.[16] The phase diagram shows that for up to 68% U, a U-Al alloy, at equilibrium, is a combination of two components: Al metal and intermetallic UAl_4. A single 13% U-Al eutectic exists in that composition range and influences the observed structure of the alloy. The hypoeutectic composition consists of Al dendrites surrounded by the eutectic. The hypereutectic composition consists of UAl_4 crystals surrounded with Al metal and the eutectic.

Given the observed alloy structure and reaction rates of Al in U-Al alloys in NaOH, it is postulated that the dissolution of the Al phase for a U-Al alloy with up to 68% U is inhibited by slower dissolution of the UAl_4 phase. In Figure 1, the convergence of the reaction rates of commercially pure Al alloys and U-Al alloys at a dissolvent concentration around 1.5 M NaOH also suggests that the Al and UAl_4 phases are simultaneously dissolving at the same rate. The reaction rate convergence also suggests there is some uranium concentration in a U-Al alloy greater than the 5% tested that will reduce the observed reaction rate to 5 to 7 mg Al/cm^2-min in boiling dissolvent at 1.5 M NaOH. That is, the Al reaction rate will become independent of the NaOH concentration in $NaOH/NaNO_3$ at a certain crossover concentration of U. At even greater U concentrations, the dissolution rate of the U-Al alloy may be reduced even further and this alloy may dissolve too slowly to be practical in a batch dissolver. Therefore, the maximum dissolution rate for Al in U-Al alloys with greater U concentrations is not expected to exceed 5 to 7 mg Al/cm^2-min.

To compensate for slower-than-expected Al dissolution rates in U-Al alloys, dissolving the alloy in high-surface-area forms is one option. Blending U-Al alloys having high and low U concentrations with pure Al and/or Al cladding to make easily dissolvable forms for dissolution is another option. If processing a U-Al alloy with a high U concentration in small amounts is desired or if this alloy produces an insoluble residue in a NaOH dissolver, batch dissolution in nitric acid is a well-established alternative process. To make the acid dissolution rates practical, Anderson and Christian[17] and other researchers recommend adding a fluoride salt catalyst to the nitric acid dissolvent.

Additional NaOH dissolution rate measurements are recommended for a U-Al alloy with the greatest U concentration expected to be dissolved. Current efforts to better characterize the expected U-Al alloys at the Department of Energy (DOE) will help define further dissolution studies.

Conclusions and Recommendations

A review of historical NaOH dissolution rates of Al metal and U-Al alloys shows a linear relationship between the reaction rate and the NaOH concentration. Specific reaction rate constants are shown to differ depending on alloy composition, dissolvent NaOH concentration, and whether $NaNO_3$ is in the dissolvent to suppress hydrogen gas production. The U-Al alloy reaction rate data from Snyder and Davis[5] correlated with the Modified Wymer model is proposed for designing U-Al alloy dissolvers. The correlation is valid for U-Al alloys with 5% to 8% U (and possibly as great as 15% U) in boiling NaOH/$NaNO_3$ dissolvent. The correlation is based on dissolvent compositions of 1.5 M to 5.5 M NaOH and 3 M $NaNO_3$ and is expected to be valid to 7.65 M NaOH or greater.

Bounding hydrogen production rates during NaOH dissolution can be calculated from the maximum Al dissolution rate and stoichiometry. In the absence of $NaNO_3$, 1.5 moles of H_2 are produced per mole of Al dissolved in NaOH. Wymer et al.[1] state that the Al dissolution rate in boiling 6 M NaOH is 36 mg/cm^2-min for a U-Al alloy with 7.5% U. Back calculation using the NaOH dissolution rate model provided by Wymer et al.[1] provides a greater aluminum dissolution rate of 42 mg/cm^2-min. Given this discrepancy, the greater dissolution rate is recommended for bounding the maximum Al dissolution and hydrogen production rates for U-Al alloys with 5% to 8% U in boiling $\leq$ 6 M NaOH. Alternatively, for pure Al (99.9 to 99.999%) and U-Al alloys containing less than 5% U, a greater bounding Al dissolution rate in boiling $\leq$ 6 M NaOH of 86 mg/cm^2-min is recommended for NaOH/$NaNO_3$ dissolvent and 151 mg/cm^2-min for NaOH dissolvent without $NaNO_3$.

U-Al alloy characterization will help define the anticipated dissolution rate range for dissolution reactor design. If U-Al alloys with greater than 13% U need to be dissolved, collection of additional rate data is recommended. Alloy structure and convergence of dissolution rates at 1.5 M NaOH suggest that at some U concentration greater than 13%, U-Al dissolution rates may decrease to 5 to 7 mg/cm^2-min or less. To compensate for this 80% reduction in dissolution rate, casting U-Al alloy into smaller forms and/or blending and casting U-Al alloy into a composition that will readily dissolve are suggested.

References

1. R. G. Wymer, D. M. Helton, A. H. Kibbey, and J. L. Land (1955), "Caustic Dissolution: Preliminary Studies," Oak Ridge National Laboratory, Oak Ridge, TN, ORNL-1782.
2. D. L. Woerner and C. Taylor (2014), "SMP Demonstrations," FY14 End Year Report PD141080, Oak Ridge Y-12 Plant, Oak Ridge, TN, Y/DZ-3742.
3. P. J. Mohr, D. B. Newell, and B. N. Taylor (2015), CODATA Recommended Values of the Fundamental Physical Constants.
4. M. E. Wieser et al. (2013), "Atomic Weights of the Elements 2011" (IUPAC Technical Report), *Pure Appl. Chem.*, 85, 5, pp. 1047-1078.
5. M. D. Snyder and M. W. Davis, Jr. (1956), "Caustic Dissolution of Uranium-Aluminum Alloy," E. I. du Pont de Nemours & Co., DP-67, Savannah River Laboratory.
6. A. T. Gresky (1952), "Recovery of Nitrogen Oxides and Rate Gas Fission Products from the Nitric Acid Dissolution of Irradiated Uranium," Oak Ridge National Laboratory, Oak Ridge, TN, ORNL-1208.
7. R. A. Charpie et al. (1954), "A Chemical Reprocessing Plant for a Nuclear Power Economy," Oak Ridge National Laboratory, Oak Ridge, TN, ORNL-1638.
8. L. B. Cunningham and C. R. Taylor (2014), "Centrifuge Evaluation for Separating the U-Al Alloy Dissolution By-Products," FY14 Report for PD130008, Oak Ridge Y-12 Plant, Oak Ridge, TN, Y/DZ-3884.
9. R. G. Wymer and R. E. Blanco (1957), "Uranium-Aluminum Alloy Dissolution," *Industrial and Engineering Chemistry*, Volume 49, No. 1.
10. D. L. Woerner (2013), "Processing Uranium Aluminum Fuel Elements," FY13 Report for PD110843, Oak Ridge Y-12 Plant, Oak Ridge, TN, Y/DZ-3595.
11. D. L. Woerner and K. K. Patton (2016), "SMP Individual Aluminum Puck Dissolution Rate Studies," Oak Ridge Y-12 Plant, Oak Ridge, TN, Y/DZ-4084.
12. R. E. Burns, H. S. Gile, and C. H. Holm (1954), "Caustic Disintegration of Uranium-Aluminum Alloy to Produce Low-Aluminum Feed for Solvent Extraction," Hanford Works, HW-31245.
13. D. R. Lide, Editor in Chief (2001), *CRC Handbook of Chemistry and Physics*, 82nd Edition, CRC Press, Boca Raton, FL.
14. The Aluminum Association (2015), "Registration Record Series Teal Sheets: International Alloy Designations and Chemical Composition Limits for Wrought Aluminum and Wrought Aluminum Alloys," The Aluminum Association, 1525 Wilson Boulevard, Arlington, VA.
15. D. E. Hamby (1951), "Fabrication of Uranium-Aluminum Alloys," Oak Ridge National Laboratory, Oak Ridge, TN, ORNL-911, p. 6.

16. T. I. Jones, I. J. McGee, and L. R. Norlock (1961), "Some Properties of Aluminum-Uranium Alloys in the Cast, Rolled and Annealed Conditions," Atomic Energy of Canada Limited, Chalk River, Ontario (Canada), AECL-1215.
17. P. A. Anderson and J. D. Christian (1998), "Alternative reagent to mercuric nitrate catalyst for dissolution of aluminum-clad nuclear fuels in nitric acid," *Journal of Materials Research*, Vol. 13, No. 1.

470583 Anode Material Studies for Electrowinning Li2O in Molten LiCl

Abstract:

Pyroprocessing has gained widespread interest for closing the nuclear fuel cycle. It is a non-aqueous, high temperature process that can be very effective at extracting actinides from spent nuclear fuel for recycle via fabrication of new reactor fuel. It was originally developed for recycle of metallic spent fuel under the Integral Fast Reactor program at Argonne National Laboratory (ANL). ANL subsequently developed an electrochemical process for reducing commercial spent oxide fuel to metals that utilizes a molten LiCl-Li_2O electrolyte. The cathode consists of a steel basket loaded with the oxide fuel. An inert anode is needed to oxidize O^{2-} ions to O_2 gas. Finding a suitable anode material has proven to be very difficult, as the conditions in the electrochemical cell are very oxidizing. Platinum wires have widely been used for this purpose, but they are incompatible with metallic lithium, which can readily form during the process. The high cost of platinum makes it economically unattractive to use pure Pt rods as disposable anodes. Two approaches have been attempted to minimize the cost of anodes in the oxide reduction process. First, titanium rods with electroplated platinum coatings have been tested as alternatives to pure Pt wire or rods. These electroplated rods provide excellent surface area per cost of platinum. In initial testing using nitrate salts containing dissolved oxides, these electrodes have proven to be durable and effective as anodes. Second, metallic lithium activity within the molten salt can be minimized by using a liquid metal such as lead as the actual cathode rather than the fuel basket. Since metallic lithium reacts with platinum, this process modification is expected to significantly extend the lifetime of the platinum coating on the anodes. In this study, Pt-coated titanium rods were used as the anode for electrowinning Li_2O to Li metal and oxygen gas. The lithium metal formed at a molten lead cathode, while the oxygen gas evolved from the surface of the platinum anodes. Post-test examination of the Pt-coated rods was used to assess the rate of corrosion of the Pt layers and estimate the long-term cost of using these rods as anodes. Successful development of these cheaper anodes is expected to greatly improve the viability of applying pyroprocessing to commercial oxide fuel.

452685 Route to Separation of Actinide Metals from Rare Earth Metals Via Conversion to Chlorides

Abstract:

Route to Separation of Actinide Metals from Rare Earth Metals via Conversion to Chlorides

Parker Okabe[1], Devin Rappleye[2], and Michael F. Simpson[1]

[1]University of Utah, Department of Metallurgical Engineering

[2]Lawrence Livermore National Laboratory

Separation of actinides from rare earths has important applications both related to rare earth metal extraction and closing the nuclear fuel cycle. Rare earths are often found in mineral deposits along with thorium, which can complicate their extraction and purification. Rare earths are also formed as fission products in spent nuclear fuel. Nuclear fuel recycling technologies aim to separate the actinides and use them as feed for fabrication of new fuel, while keeping the actinides out of the waste streams that are destined for permanent disposal in geologic repositories. Pyroprocessing of spent fuel can be used to electrochemically separate actinides from rare earths and generate rare earth-rich metallic waste forms. But the proximity of the reduction potentials for actinides and rare earths make it very difficult to eliminate carryover of actinides into the rare earth waste forms. The rare earth waste forms will be much simpler and cheaper to dispose if the actinides can be completely removed. It is, thus, important to develop a process to reduce actinide contamination in rare earth metals. While aqueous separations methods involving acid dissolution followed by extraction into an organic phase have been scaled up and implemented commercially in several countries, this approach is widely viewed as unattractive due to actinide proliferation and environmental concerns. There is, thus, increasing interest in pyrochemical separations approaches that eliminate liquid wastes and demonstrate low selectivity for actinide separation. One such pyrochemical approach that we are investigating is based on physical property differences between actinide and rare earth chlorides. In order to develop this approach, experiments have been run with cerium and cerium+uranium metal. Metallic samples were made by melting cerium and uranium together into a small bead. In a batch reactor, these metal beads were first hydrided with pure H_2 to form metal hydride powders and then were chlorinated with anhydrous chlorine gas. Pressure readings were used to monitor the extent of reaction versus time. Hydriding was pursued to reduce the particle size and facilitate rapid and complete chlorination. Complete hydriding could be accomplished in 3 hrs at 300°C, provided the starting size of the metal beads was approximately 5 mm in diameter or smaller. Otherwise, unreacted metal cores were present.

This work performed under the auspices of the U.S. Department of Energy for Lawrence Livermore National Laboratory under Contract DE-AC52-07NA27344

469409 A High-Throughput Investigation of Fe-Cr-Al As a Novel High-Temperature Coating for Nuclear Cladding Applications

Abstract:

During a reactor cooling system failure in a nuclear power plant, the reactor temperature rapidly rises. This results in the production of high-temperature and high-pressure steam within the reactor, which is periodically vented to prevent a failure of the nuclear containment vessel. At temperatures exceeding 1473 K, current generation Zircalloy claddings can undergo multiple structural phase transformations or react with steam to form ZrO_2 and hydrogen gas. The explosion of hydrogen gas during venting was one of the key contributing factors to the magnitude of the Fukushima disaster, as it damaged multiple containment vessels for units 1 and 3. A number of recent metallurgical studies have focused on the validation of Fe-Cr-Al alloys as possible replacements for Zircalloy due to its superior resistance to oxidation in steam environments. Unfortunately, the certification process for new cladding materials is time consuming. Conversely, coating technologies for existing cladding materials could potentially be more rapidly deployed while permitting for full replacements is completed. However, it is well known that deposited coatings can exhibit substantially different oxidation properties from their bulk counterparts. The coating composition is particularly problematic as subatomic percent variation in Al content can mark the difference between forming a passivating versus non-passivating oxide.

Here, I will describe the use of a MGI-based framework to investigate the role of synthesis and processing on Fe-Cr-Al thin films. Using computationally guided synthesis, high-throughput structural characterization and large scale data analytics we have systematically investigated oxidation resistance for the region of the phase diagram identified by the bulk community to resist oxidation. A primary screen, consisting of in-situ glancing incidence synchrotron diffraction during air oxidation, revealed that alloys containing greater than 3.08 at% Al and between 20.0 at% and 32.9at% Cr showed reduced formation of non-passivating oxides. Subsequent secondary screens revealed the formation of a

passivating oxide for time periods in excess of 6 hours.

472428 Development of a Rotating Cylinder Electrode for Electroanalytical Measurements in a Used Nuclear Fuel Electrorefiner

Abstract:

Pyroprocessing is an important application of metallurgy to nuclear energy systems as an attractive option for the recycling of actinides from used nuclear fuel. This technology utilizes differences in the electrochemical potentials of metal ions to selectively separate actinides from lanthanides and other fission products found in used fuel. The ability to identify the co-deposition of uranium with other ions (actinides or contaminates) and determine the rate of deposition for each metal would be beneficial to process control and safeguards. DREP (Deposition Rates from Electrode Potentials) was developed to predict the metal(s) depositing at the cathode of an electrorefiner. DREP is an electrorefining model that couples the cell current and cathode potential to determine if multiple metals are depositing and, if so, predict the rate of deposition for each metal. A key assumption to DREP is a constant, well-defined diffusion layer. The diffusion layer at a stationary electrode grows with time until natural convection prevents further growth. Rotating electrode can provide a well-defined, time-independent diffusion layer. Rotating disk (RDE) and rotating ring-disk electrodes (RRDE) have been developed for application in various high-temperature molten salts. However, the RDE and RRDE rely upon an insulator sealing the edges of an electrode to prevent leakage of electrolyte into the electrode-insulator interface. This leakage is exacerbated at high-temperature. Furthermore, many insulators react with molten salts degrading the insulating and/or sealing properties of the selected material. The rotating cylinder electrode (RCE) can circumvent many of these challenges since no insulating material is required. Since no previous work on a molten salt RCE could be found in the literature, a RCE for electrochemical measurements in molten LiCl-KCl eutectic was developed. Electroanalytical measurements made with RCE are also presented. The theory developed for the RCE in aqueous solutions is applied to the electroanalytical measurements in molten LiCl-KCl eutectic to determine its validity in molten salts and to evaluate the experimental setup. Lastly, short-term (<1 min.) electrodeposition test were performed in molten LiCl-KCl eutectic containing UCl_3 and $MgCl_2$ (surrogate for $PuCl_3$)

in which both uranium and magnesium were deposited at the cathode. The tests were used to validate DREP and evaluate its accuracy in predicting uranium and magnesium co-deposition.

AUTHOR INDEX

AIChE
120 Wall Street, FL 23
New York, NY 10005-4020

CURRAN ASSOCIATES INC.
proceedings.com

ISBN 978-1-5108-3443-9